Lecture Notes in Computer Science 4436

Commenced Publication in 1973
Founding and Former Series Editors:
Gerhard Goos, Juris Hartmanis, and Jan va'

Christopher R. Stephens   Marc Toussaint
Darrell Whitley   Peter F. Stadler (Eds.)

Foundations of Genetic Algorithms

9th International Workshop, FOGA 2007
Mexico City, Mexico, January 8-11, 2007
Revised Selected Papers

 Springer

Volume Editors

Christopher R. Stephens
Universidad Nacional Autonoma de Mexico, Instituto de Ciencias Nucleares
Circuito Exterior, A. Postal 70-543, Mexico D.F. 04510, Mexico
E-mail: stephens@nucleares.unam.mx

Marc Toussaint
TU Berlin
Franklinstr. 28/29, 10587 Berlin, Germany
E-mail: mtoussai@cs.tu-berlin.de

Darrell Whitley
Colorado State University, Department of Computer Science
Fort Collins, CO 80523, USA
E-mail: whitley@cs.colostate.edu

Peter F. Stadler
Universität Leipzig, Institut für Informatik
Härtelstr. 16-18, 04107 Leipzig, Germany
E-mail: studla@bioinf.uni-leipzig.de

Library of Congress Control Number: 2007929644

CR Subject Classification (1998): F.1-2, I.2, I.2.6, I.2.8, D.2.2

LNCS Sublibrary: SL 1 – Theoretical Computer Science and General Issues

ISSN   0302-9743
ISBN-10   3-540-73479-1 Springer Berlin Heidelberg New York
ISBN-13   978-3-540-73479-6 Springer Berlin Heidelberg New York

Springer is a part of Springer Science+Business Media

springer.com

© Springer-Verlag Berlin Heidelberg 2007
Printed in Germany

Typesetting: Camera-ready by author, data conversion by Scientific Publishing Services, Chennai, India
Printed on acid-free paper   SPIN: 12087204   06/3180   5 4 3 2 1 0

Editorial Introduction

Since their inception in 1990, the FOGA (Foundations of Genetic Algorithms) workshops have been one of the principal reference sources for theoretical developments in evolutionary computation (EC) and, in particular, genetic algorithms (GAs). The ninth such workshop, FOGA IX, was held at the Instituto de Ciencias Nucleares of the Universidad Nacional Autónoma de México, Mexico City during January 8–11, 2007.

One of the main reasons the FOGA series of conferences has had a large impact in EC has been its distinct profile as the only conference dedicated to theoretical issues of a "foundational" nature – both conceptual and technical. In this FOGA conference, and in keeping with this tradition, special attention was paid to the biological foundations of EC. The essential mathematical structure behind many evolutionary algorithms is the one familiar from population genetics, whose basic elements have been around now for at least 70 years. The last 20 years or so, however, have witnessed huge changes in our understanding of how genomes and other genetic structures work due to a plethora of new experimental techniques and results. How does this new phenomenology change our understanding of what genetic systems do and how they do it? And how can we design "better" ones?

In this spirit, the first 2 days of the conference consisted of organized discussions built around sets of lectures given by two world authorities on the "old" biology and the "new" biology – Reinhard Burger (University of Vienna) and Jim Shapiro (University of Chicago). The motivation behind this was that by a careful presentation of the main ideas, a useful transfer of knowledge of the latest developments and understanding of genetic dynamics in biology would be fruitful for the EC community in better understanding and designing artificial genetic systems. In particular the following questions were addressed:

- How do real genetic systems work?
- Why do they work that way?
- From this, what can we learn in order to design "better" artificial genetic systems?

One of the most important conclusions from this confrontation between the old and the new, was that the genotype – phenotype map and the huge variety of complex ways by which genomes can interchange and mix genetic material are not represented adequately in the standard "selection on a fixed fitness landscape, mutation and homologous recombination" picture so dominant in EC and, particularly, GAs. Secondly, it became clear that the canonical picture of population genetics was not an appropriate framework for considering "macroevolution" over long time scales, where the restructuring of genomes can be enormous. Both these facts potentially pose great challenges for EC. For instance, under what circumstances are all the diverse exchange and restructuring mechanisms for genomes useful in an EC setting? It is hard to imagine that optimizing the 3,456-city Travelling Salesman problem needs such sophisticated apparatus. Such a limited combinatorial optimization context is probably much more akin to the evolution of specific phenotypic characteristics, as treated in standard population genetics. No doubt that is one of the main reasons for the success of GAs in combinatorial optimization. However, it is not clear if such a paradigm is adequate for producing a more intelligent robot.

To understand then why biology uses certain representations and operators, it is necessary to understand what a biological system has to "do" when compared with EC systems. Surviving in an uncertain, time-dependent environment is surely an infinitely more complex task than finding a set of allele values that represent an optimal solution to a combinatorial optimization problem. In this sense, one may wonder if there are any biological systems that are at least similar to typical problems faced in EC. Peter Stadler presented probably one of the closest analogies – evolution of macromolecules in the context of an RNA world – where the fitness function for a particular RNA configuration is its replication rate. However, such simple chemical evolution seems far removed from the macro-evolution of entire organisms. Hopefully, some of the fruits of this more intense examination of the relationship between biological evolution and EC will appear in the next FOGA.

The second two days of the conference were of a more standard FOGA format with contributed talks and ample time for discussion between them. For this workshop there were 22 submissions which were each sent in a double-blind review to three referees. Twelve high quality submissions that cover a wide range of theoretical topics were eventually accepted after two more rounds of revisions and are presented in this volume.

We would like to thank our co-organizers, Peter Stadler and Darrell Whitley, for their efforts and input. Katya Rodríguez formed part of the Local Organizing Committee and played an important role in making the conference run smoothly, as did Trinidad Ramírez and various student helpers. Thanks go to the Instituto de Ciencias Nucleares for providing its facilities and to the Macroproyecto Tecnologias para la Universidad de la Información y de la Computación for financial and technical support.

April 2007   Christopher R. Stephens
                                                           Marc Toussaint

Organization

FOGA 2007 was organized in cooperation with ACM/SIGEVO at the Instituto de Ciencias Nucleares, Universidad Nacional Autonoma de Mexico (UNAM), Mexico City, January 8–11, 2007.

Executive Committees

Organizing Committee:
    Chris Stephens (UNAM)
    Darrell Whitley (Colorado State University)
    Peter Stadler (University of Leipzig)
    Marc Toussaint (University of Edinburgh)

Local Organizing Committee:
    Chris Stephens (UNAM)
    Katya Rodriguez (UNAM)

Program Committee:
    Chris Stephens (UNAM)
    Darrell Whitley (Colorado State University)
    Peter Stadler (University of Leipzig)
    Marc Toussaint (University of Edinburgh)

Referees

J.E. Rowe	A. Eremeev	M. Schoenauer
W.B. Langdon	R. Heckendorn	J. Smith
A. Prügel-Bennett	A.H. Wright	T. Jansen
C. Witt	H.-G. Beyer	R. Poli
R. Drechsler	M. Pelikan	W. Gutjahr
J. Branke	Y. Gao	A. Auger
W.E. Hart	M. Gallagher	P. Stadler
C. Igel	J. Shapiro	D. Whitley
L.M. Schmitt	J. He	M. Vose
B. Mitavskiy	S. Droste	O. Teytaud
I. Wegener	A. Bucci	

Sponsoring Institutions

ACM Special Interest Group on Genetic and Evolutionary Computation, SIGEVO.
Instituto de Ciencias Nucleares, UNAM.
Macroproyecto Universitario "Tecnologias para la Universidad de la Informacion y la Computacion," UNAM.
Posgrado en Ciencia y Ingenieria de la Computacion, UNAM.

Table of Contents

Inbreeding Properties of Geometric Crossover and Non-geometric
Recombinations... 1
  Alberto Moraglio and Riccardo Poli

Just What Are Building Blocks? 15
  Christopher R. Stephens and Jorge Cervantes

Sufficient Conditions for Coarse-Graining Evolutionary Dynamics 35
  Keki Burjorjee

On the Brittleness of Evolutionary Algorithms....................... 54
  Thomas Jansen

Mutative Self-adaptation on the Sharp and Parabolic Ridge 70
  Silja Meyer-Nieberg and Hans-Georg Beyer

Genericity of the Fixed Point Set for the Infinite Population Genetic
Algorithm.. 97
  Tomáš Gedeon, Christina Hayes, and Richard Swanson

Neighborhood Graphs and Symmetric Genetic Operators 110
  Jonathan E. Rowe, Michael D. Vose, and Alden H. Wright

Decomposition of Fitness Functions in Random Heuristic Search 123
  Yossi Borenstein and Riccardo Poli

On the Effects of Bit-Wise Neutrality on Fitness Distance Correlation,
Phenotypic Mutation Rates and Problem Hardness.................... 138
  Riccardo Poli and Edgar Galván-López

Continuous Optimisation Theory Made Easy? Finite-Element
Models of Evolutionary Strategies, Genetic Algorithms and
Particle Swarm Optimizers 165
  *Riccardo Poli, William B. Langdon, Maurice Clerc, and
  Christopher R. Stephens*

Saddles and Barrier in Landscapes of Generalized Search Operators 194
  *Christoph Flamm, Ivo L. Hofacker, Bärbel M.R. Stadler, and
  Peter F. Stadler*

Author Index... 213

Inbreeding Properties of Geometric Crossover and Non-geometric Recombinations

Alberto Moraglio and Riccardo Poli

Department of Computer Science, University of Essex, UK
{amoragn,rpoli}@essex.ac.uk

Abstract. Geometric crossover is a representation-independent generalization of traditional crossover for binary strings. It is defined in a simple geometric way by using the distance associated with the search space. Many interesting recombination operators for the most frequently used representations are geometric crossovers under some suitable distance. Showing that a given recombination operator is a geometric crossover requires finding a distance for which offspring are in the metric segment between parents. However, proving that a recombination operator is not a geometric crossover requires excluding that one such distance exists. It is, therefore, very difficult to draw a clear-cut line between geometric crossovers and non-geometric crossovers. In this paper we develop some theoretical tools to solve this problem and we prove that some well-known operators are not geometric. Finally, we discuss the implications of these results.

1 Introduction

A fitness landscape [23] can be visualised as the plot of a function resembling a geographic landscape, when the problem representation is a real vector. When dealing with binary strings and other more complicated combinatorial objects, e.g., permutations, however, the fitness landscape is better represented as a height function over the nodes of a simple graph [19], where nodes represent locations (solutions), and edges represent the relation of direct neighbourhood between solutions.

An abstraction of the notion of landscape encompassing all the previous cases is possible. The solution space is seen as a metric space and the landscape as a height function over the metric space [1]. A metric space is a set endowed with a notion of distance between elements fulfilling few axioms [3]. Specific spaces have specific distances that fulfil the metric axioms. The ordinary notion of distance associated with real vectors is the Euclidean distance, though there are other options, e.g., Minkowski distances. The distance associated to combinatorial objects is normally the length of the shortest path between two nodes in the associated neighbourhood graph [4]. For binary strings, this corresponds to the Hamming distance.

In general, there may be more than one neighbourhood graph associated to the same representation, simply because there can be more than one meaningful

C.R. Stephens et al. (Eds.): FOGA 2007, LNCS 4436, pp. 1–14, 2007.

notion of syntactic similarity applicable to that representation [10]. For example, in the case of permutations, the adjacent element swap distance and the block reversal distance are equally natural notions of distance. Different notions of similarity are possible because the same permutation (genotype) can be used to represent different types of solutions (phenotypes). For example, permutations can represent solutions of a problem where relative order is important. However, they can also be used to represent tours, where the adjacency relationship among elements is what matters [21].

The notion of fitness landscape is useful if the search operators employed are connected or matched with the landscape: the stronger the connection the more landscape properties mirror search properties. Therefore, the landscape can be seen as a function of the search operator employed [5]. Whereas mutation is intuitively associated with the neighbourhood structure of the search space, crossover stretches the notion of landscape leading to search spaces defined over complicated topological structures [5].

Geometric crossover and geometric mutation [9] are representation-independent search operators that generalise by abstraction many pre-existing search operators for the main representations used in EAs, such as binary strings, real vectors, permutations and syntactic trees. They are defined in geometric terms using the notions of line segment and ball. These notions and the corresponding genetic operators are well-defined once a notion of distance in the search space is defined. This way of defining search operators as function of the search space is the opposite to the standard approach in which the search space is seen as a function of the search operators employed. Our new point of view greatly simplifies the relationship between search operators and fitness landscape and allows different search operators to share the same search space.

The reminder of this paper is organized as follows. In section 2, we introduce the geometric framework. In section 3, we show that the definition of geometric crossover can be cast in two equivalent, but conceptually very different, forms: functional and existential. When proving geometricity the existential form is the relevant one. We use this form also to show why proving non-geometricity of an operator looks impossible. In section 4, we develop some general tools to prove non-geometricity of recombination operators. In section 5, we prove that three recombination operators for vectors of reals, permutations and syntactic trees representations are not geometric. Importantly this implies that there are two *non-empty* representation-independent classes of recombination operators: geometric crossovers and non-geometric crossovers. In section 6, we draw some conclusions and present future work.

2   Geometric Framework

2.1   Geometric Preliminaries

In the following we give necessary preliminary geometric definitions and extend those introduced in [9]. For more details on these definitions see [4].

The terms *distance* and *metric* denote any real valued function that conforms to the axioms of identity, symmetry and triangular inequality. A simple connected graph is naturally associated to a metric space via its *path metric*: the distance between two nodes in the graph is the length of a shortest path between the nodes. Distances arising from graphs via their path metric are called *graphic distances*. Similarly, an edge-weighted graph with strictly positive weights is naturally associated to a metric space via a *weighted path metric*.

In a metric space (S, d) a *closed ball* is a set of the form $B_d(x; r) = \{y \in S | d(x, y) \leq r\}$ where $x \in S$ and r is a positive real number called the radius of the ball. A *line segment* (or closed interval) is a set of the form $[x; y]_d = \{z \in S | d(x, z) + d(z, y) = d(x, y)\}$ where $x, y \in S$ are called extremes of the segment. Metric ball and metric segment generalize the familiar notions of ball and segment in the Euclidean space to any metric space through distance redefinition. These generalized objects look quite different under different metrics. Notice that the notions of metric segment and shortest path connecting its extremes (*geodesic*) do not coincide as it happens in the specific case of an Euclidean space. In general, there may be more than one geodesic connecting two extremes; the metric segment is the union of all geodesics.

We assign a structure to the solution set S by endowing it with a notion of distance d. $M = (S, d)$ is therefore a solution *space* (or search space) and $L = (M, g)$ is the corresponding *fitness landscape* where $g : S \to \mathbb{R}$ is the fitness function. Notice that in principle d could be arbitrary and need not have any particular connection or affinity with the search problem at hand.

2.2 Geometric Crossover Definition

The following definitions are *representation-independent* and, therefore, crossover is well-defined for any representation. Being based on the notion of metric segment, *crossover is only function of the metric d* associated with the search space.

A recombination operator OP takes parents p_1, p_2 and produces one offspring c according to a given conditional probability distribution:

$$Pr\{OP(p_1, p_2) = c\} = Pr\{OP = c | P_1 = p_1, P_2 = p_2\} = f_{OP}(c | p_1, p_2)$$

Definition 1 (*Image set*). *The* image set $Im[OP(p_1, p_2)]$ *of a genetic operator OP is the set of all possible offspring produced by OP with non-zero probability when parents are p_1 and p_2.*

Definition 2 (*Geometric crossover*). *A recombination operator CX is a geometric crossover under the metric d if all offspring are in the segment between its parents:* $\forall p_1, p_2 \in S : Im[CX(p_1, p_2)] \subseteq [p_1, p_2]_d$

Definition 3 (*Uniform geometric crossover*). *The uniform geometric crossover UX under d is a geometric crossover under d where all z laying between parents x and y have the same probability of being the offspring:*

$$\forall x, y \in S : f_{UX}(z|x, y) = \frac{\delta(z \in [x; y]_d)}{|[x; y]_d|}$$

$$Im[UX(x, y)] = \{z \in S | f_{UX}(z|x, y) > 0\} = [x; y]_d$$

where δ is a function that returns 1 if the argument is true, 0 otherwise.

A number of general properties for geometric crossover and mutation have been derived in [9].

2.3 Notable Geometric Crossovers

For vectors of reals, various types of blend or line crossovers, box recombinations, and discrete recombinations are geometric crossovers [9]. For binary and multary strings (fixed-length strings based on a n symbols alphabet), all mask-based crossovers (one point, two points, n-points, uniform) are geometric crossovers [9,13]. For permutations, PMX, Cycle crossover, merge crossover and others are geometric crossovers [10,11]. For Syntactic trees, the family of Homologous crossovers (one-point, uniform crossover) are geometric crossovers [12]. Recombinations for other more complicated representations such as variable length sequences, graphs, permutations with repetitions, circular permutations, sets, multisets partitions are geometric crossovers [15,9,10,14].

2.4 Geometric Crossover Landscape

Since our geometric operators are representation-independent, one might wonder as to the usefulness of the notion of geometricity and geometric crossovers in practical applications. To see this, it is important to understand the difference between problem and landscape.

Geometric operators are defined as functions of the distance associated to the search space. However, the search space does not come with the problem itself. The problem consists only of a fitness function to optimize, that defines what a solution is and how to evaluate it, but it does not give any structure over the solution set. The act of putting a structure over the solution set is part of the search algorithm design and it is a designer's choice. A fitness landscape is the fitness function plus a structure over the solution space. So, for each problem, there is one fitness function but as many fitness landscapes as the number of possible different structures over the solution set. In principle, the designer could choose the structure to assign to the solution set completely independently from the problem at hand. However, because the search operators are defined over such a structure, doing so would make them decoupled from the problem, hence turning the search into something very close to random search.

In order to avoid this one can exploit problem knowledge in the search. This can be achieved by carefully designing the connectivity structure of the fitness landscape. That is, the landscape can be seen as a knowledge interface between algorithm and problem [10]. In [10] we discussed three heuristics to design the connectivity of the landscape in such a way to aid the evolutionary search performed by geometric crossover. These are: i) pick a crossover associated to a good mutation, ii) build a crossover using a neighbourhood structure based on the small-move/small-fitness-change principle, and iii) build a crossover using a distance that is relevant for the solution interpretation.

Once the connectivity of the landscape is correctly designed, problem knowledge can be exploited by search operators to perform better than random search, even if the search operators are problem-independent (as in the case of geometric crossover and mutation). Indeed, by using these heuristics, we have *designed* very effective geometric crossovers for N-queens problem [11], TSP [11] [10], Job Shop Scheduling [11], Protein Motifs discovery [20], Graph Partitioning [6], Sudoku [16] and Finite State Machines [7].

3   Interpretations of the Definition of Geometric Crossover

In section 2, we have defined geometric crossover as function of the distance d of the search space. In this section we take a closer look at the meaning of this definition *when the distance d is not known*. We identify three fundamentally different interpretations of the definition of geometric crossover. Interestingly it will become evident that there is an inherent element of self-reference in the definition. We show that proving that a recombination operator is non-geometric may be impossible.

3.1   Functional Interpretation

Geometric crossover is function of a *generic distance*. If one considers a specific distance one can obtain a specific geometric crossover for that distance by functional application of the definition of geometric crossover to this distance. This approach is particularly useful when the specific distance is firmly rooted in a solution representation (e.g., edit distances). In this case, in fact, the specification of the definition of geometric crossover to the distance acts as a formal recipe that indicates how to manipulate the syntax of the representation to produce offspring from parents. This is a general and powerful way to get new geometric crossover for any type of solution representation. For example, given the Hamming distance on binary string by functional application of the definition of geometric crossover we obtain the family of mask-based crossover for binary strings. In particular, by functional application of the definition of uniform geometric crossover one obtains the traditional uniform crossover for binary strings.

3.2   Abstract Interpretation

The second use of the definition of geometric crossover does not require to specify any distance. In fact we do apply the definition of geometric crossover to a generic distance. Since the distance is a metric that is a mathematical object defined axiomatically, the definition of geometric crossover becomes an axiomatic object as well. This way of looking at the definition of geometric crossover is particularly useful when one is interested in deriving general theoretical results that hold for geometric crossover under any specific metric. We will use this abstract interpretation in section 4 to prove the inbreeding properties that are common to all geometric crossovers.

3.3 Existential Interpretation

The third way of looking at the definition of geometric crossover becomes apparent when the distance d is not known and we want to find it. This happens when we want to know whether a recombination operator RX, defined operationally as some syntactic manipulation on a specific representation, is a geometric crossover and for what distance. This question hides an element of self-reference of the definition of geometric crossover. In fact what we are actually asking is: given that *the geometric crossover is defined over the metric space it induces by manipulating the candidate solutions*, what is such a metric space for RX if any?

The self-reference arises from the fact that the definition of geometric crossover applies at two distinct levels at the same time: (a) at a representation level, as a manipulation of candidate solutions, and (b) at a geometric level, on the underlying metric space based on a geometric relation between points. This highlights the inherent *duality* between these two worlds: they are based on the *same* search space seen from opposite viewpoints, from the representation side and from the metric side.

Self-referential statements can lead to paradoxes. Since the relation between geometric crossover and search space is what ultimately gives it all its advantages, it is of fundamental importance to make sure that this relation sits on a firm ground. So, it is important to show that the definition of geometric crossover does not lead to any paradox. We show in the following that the element of self-reference can be removed and the definition of geometric crossover can be cast in existential terms making it paradox-free.

A non-functional definition of geometric crossover is the following: a recombination operator RX is a geometric crossover if the induced search space is a metric space on which RX can be defined as geometric crossover using the functional definition of geometric crossover. This is a self-referential definition. If a recombination operator does not induce any metric space on which it can be defined as geometric crossover, then it is a non-geometric crossover.

We can remove the element of self-reference from the previous definition and cast it in an existential form: a recombination RX is a geometric crossover if for any choice of the parents all the offspring are in the metric segment between them for some metric.

The existential definition is equivalent to the self-referential definition because if such a metric exists the operator RX can be defined as geometric crossover on such a space. On the other hand, if an operator is defined on a metric space as geometric crossover in a functional form, such a space exists by hypothesis and offspring are in the segment between parents under this metric by definition.

3.4   Geometric Crossover Classes

The functional definition of geometric crossover induces a natural existential classification of all recombination operators into two classes of operators:

- *geometric crossover class* $\mathcal{G}$: a recombination OP belongs to this class if there exists at least a distance d under which such a recombination is geometric: $OP \in \mathcal{G} \iff \exists d : \forall p_1, p_2 \in S : Im[OP(p_1, p_2)] \subseteq [p_1, p_2]_d$.
- *non-geometric crossover class* $\bar{\mathcal{G}}$: a recombination OP belongs to $\bar{\mathcal{G}}$ if there is no distance d under which such a recombination is geometric: $OP \in \bar{\mathcal{G}} \iff \forall d : \exists p_1, p_2 \in S : Im[OP(p_1, p_2)] \setminus [p_1, p_2]_d \neq \emptyset$.

For this classification to be meaningful we need these two classes to be non-empty. In previous work we proved that a number of recombination operators are geometric crossovers so $\mathcal{G}$ is not empty. What about $\bar{\mathcal{G}}$? To prove that this class is not empty we have to prove that at least one recombination operator is non-geometric. However, as we illustrate below this is not easy to do.

Let us first illustrate how one can prove that a recombination operator RX is in $\mathcal{G}$. We will use the self-referential definition of geometric crossover. The procedure is the following: guess a candidate distance d, then prove that all offspring of all possible pairs of parents are in the metric segment associated with d. If this is true then the recombination RX is geometric crossover under the distance d *because the operator RX can be defined as a geometric crossover on this space*. If the distribution of the offspring in the metric segments under d is uniform, RX is the uniform geometric crossover for the metric d *because the operator RX can be defined as the (unique) geometric uniform crossover on this space*. If one finds that some offspring are not in the metric segment between parents under the initially guessed distance d then the operator RX cannot be defined as geometric crossover over this space. However, this does not imply $RX \in \bar{\mathcal{G}}$ because there may exist another metric d' that fits RX and *makes it definable* as a geometric crossover on d'. So, one has to guess a new candidate distance for RX and start all over again until a suitable distance is found.

Although we developed some heuristics for the selection of a candidate distance, in general proving that a recombination operator is geometric may be quite hard (see for example [12] where we considered homologous crossover for GP trees). Nonetheless, the approach works and, in previous work, we proved that a number of recombination operators for the most frequently used representations are geometric crossover under suitable distances.

It is evident, however, that the procedure just described cannot be used to prove that a given recombination operator RX is non-geometric. This is because we would need to test and exclude all possible distances, which are infinitely many, before being certain that RX is not geometric. Clearly, this is not possible.

In the next section we build some theoretical tools based on the abstract interpretation of the definition of geometric crossover to prove non-geometricity in a more straightforward way.

4   Inbreeding Properties of Geometric Crossover

How could we actually prove non-geometricity? From the definition of geometric crossover based on a generic notion of distance (abstract interpretation, see section 3.2), we could derive metric properties that are common to the class of all geometric crossovers and that could be tested without making explicit use of the distance.

Any reference to the distance needs necessarily to be excluded from these properties because what in fact we need to test is the existence of an underlying distance behind a given recombination operator hence we cannot assume the existence of one a priori. So the first requirement is that these properties derive from the metric axioms but cannot be about distance. A second requirement is generality: these properties need to be representation-independent so that recombination for any solution representation can be tested. A third and last requirement is that these properties need to be independent from the specific probability distribution with which offspring are drawn from the segment between the parents. In particular they must encompass also geometric crossovers where offspring are drawn from only part of the segment.

If necessary properties satisfying these requirements existed, testing a recombination operator for non-geometricity would become straightforward: if such operator does not have a property common to all geometric crossovers it is automatically non-geometric. Fortunately, properties of this type do exists. They are the inbreeding properties of geometric crossover.

In the following we introduce three fundamental properties of geometric crossover arising only from its axiomatic definition (metric axioms), hence valid for any distance, any probability distribution and any underlying solution representation. These properties of geometric crossover are simple properties of geometric interval spaces [22] adapted to the geometric crossover. The properties proposed are based on inbreeding (breeding between close relatives) using geometric crossover and avoid explicit reference to the solution representation. In section 5, we will make good use of these properties to prove some non-geometricity results.

Theorem 1 (*Property of Purity*). *If the operator RX is geometric then the recombination of one parent with itself can only produce the parent itself.*

Proof: If RX is geometric there exists a metric d such that any offspring o belongs to the segment between parents s_1, s_2 under metric d: $d(s_1, o) + d(o, s_2) = d(s_1, s_2)$. When the parents coincide, $s = s_1 = s_2$, we have: $d(s, o) + d(o, s) = d(s, s)$ hence for symmetry and identity axioms of metric $d(s, o) = 0$ for any metric. For the identity axiom this implies $o = s$.   □

Inbreeding diagram of the property of purity (see Fig. 1(a)): when the two parents are the same $P1$, their child C must be $P1$.

Theorem 2. (*Property of Convergence*) *If the operator RX is geometric then the recombination of one parent with one offspring cannot produce the other parent of that offspring unless the offspring and the second parent coincide.*

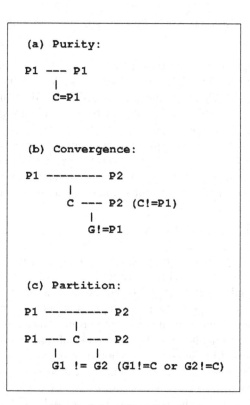

Fig. 1. Inbreeding diagrams

Proof: If RX is geometric there exists a metric d such that for any offspring o of parents s_1 and s_2 we have $d(s_1, o) + d(o, s_2) = d(s_1, s_2)$. If one can produce parent s_2 by recombining s_1 and o, it must be also true that $d(s_1, o) = d(s_1, s_2) + d(s_2, o)$. By substituting this last expression in the former one we have: $d(s_1, s_2) + d(s_2, o) + d(o, s_2) = d(s_1, s_2)$, which implies $d(o, s_2) = 0$ and $s_2 = o$ for any metric. □

Inbreeding diagram of the property of convergence (see Fig. 1(b)): two parents $P1$ and $P2$ produce the child C. We consider a C that does not coincide with $P1$. The child C and its parent $P2$ mate and produce a grandchild G. The property of convergence states that G can never coincide with $P1$.

Theorem 3. *(Property of Partition) If the operator RX is geometric and c is a child of a and b, then the recombination of a with c and the recombination of b with c cannot produce a common grandchild e other than c.*

Proof: We have that $c \in [a, b]$, $e \in [a, c]$ and $e \in [b, c]$, from which it follows that $d(a, c) + d(c, b) = d(a, b)$, $d(a, e) + d(e, c) = d(a, c)$ and $d(b, e) + d(e, c) = d(b, c)$. Substituting the last two expressions in the first one we obtain:

$$d(a, e) + d(e, c) + d(b, e) + d(e, c) = d(a, b)$$

Notice that $d(a, e) + d(b, e) \geq d(a, b)$ and, so, the previous equation implies $d(e, c) = 0$ and $e = c$. □

Inbreeding diagram of the property of partition (see Fig. 1(c)): two parents $P1$ and $P2$ produce the child C. The child C mates with both its parents, $P1$ and $P2$, producing grandchildren $G1$ and $G2$, respectively. We consider the case in which at least one grandchildren is different from C. The property of partition states that $G1$ and $G2$ can never coincide.

Geometric crossovers whose offspring cover completely the segments between their parents (complete geometric crossovers) have a larger set of properties including extensiveness ($a, b \in Im(UX(a, b))$) and symmetry ($Im(UX(a, b)) = Im(UX(b, a))$), which however, are not common to all geometric crossovers.

4.1   Relation with Forma Analysis

Since the inbreeding properties of geometric crossover are related with forma analysis [18] we briefly explain this relation.

Radcliffe developed a theory [18] of recombination operators starting from the notion of forma that is a representation-independent generalization of schema. A forma is an equivalence class on the space of chromosomes induced by a certain equivalence relation. Radcliffe describes a number of important formal *desirable properties* that a recombination operator should respect to be a good recombination operator. These properties are representation-independent and are stated as requirements on how formae should be manipulated by recombination operators.

Geometric crossover, on the other hand, is formally defined geometrically using the distance associated with the search space. Unlike Radcliffe's properties, the inbreeding properties of geometric crossover are not desired properties but are properties that are *common* to all geometric crossovers and derive logically from its formal definition only.

It is important to highlight that geometric crossover theory and forma analysis overlap but they are not isomorphic. This becomes clear when we consider what schemata for geometric crossover are. In forma theory, the recombination operators introduced by Radcliffe "respect" formae: offspring must belong to the same formae both parents belong to. A natural generalization of schemata for geometric crossover in this sense are (metric) convex sets: offspring in the line segment between parents belong to all convex sets common to their parents. So *geometric crossover induces a convexity structure over the search space.* A convexity structure is not the same thing as an equivalence relation: convex sets, like equivalence classes, cover the entire space but unlike them convex sets do not partition the search space because they overlap. Interestingly, convex sets seen as schemata naturally unify the notions of inheritance and fitness landscape.

A further advantage of geometric crossover over forma theory is that whereas it is rather easy to define and deal with distances for complex representations such as trees and graphs (using edit distances) it is much harder to use equivalence classes.

5 Non-geometric Crossovers

In the following we use the properties of purity, convergence and partition to prove the non-geometricity of three important recombination operators: extended line recombination, Koza's subtree swap crossover and Davis's Order Crossover (see, for example, [2] for a description of these operators).

Theorem 4. *Extended line recombination is not a geometric crossover.*

Proof: The convergence property fails to hold. Let p_1 and p_2 be two parents, and o the offspring lying in the extension line beyond p_1. It is easy to see that using the extension line recombination on o and p_2, one can obtain p_1 as offspring. $\square$

Theorem 5. *Koza's subtree swap crossover is not a geometric crossover.*

Proof: The property of purity fails to hold. Subtree swap crossover applied to two copies of the same parent may produce offspring trees different from it. $\square$

Theorem 6. *Davis's Order Crossover is non-geometric.*

Proof: The *convergence property does not hold* in the counterexample in Figure 2 where the last offspring coincides with parent 2. $\square$

What are the implications of knowing that these operators are *not* geometric? The first one is that one is not tempted to try to prove its geometricity with yet another distance.

A second immediate and fundamental consequence of knowing that an operator is non-geometric is that since it is not associable with any metric it is

```
Parent 1 : 12.34.567
Parent 2 : 34.56.127
Section  : --.34.---
Available elements in order: 12756

Offspring: 65.34.127
Parent 3 := Offspring

Parent 3 : 6534.12.7
Parent 1 : 1234.56.7
Section  : ----.12.-
Available elements in order: 73456

Offspring: 3456.12.7
Offspring = Parent 2
```

Fig. 2. Counterexample to the geometricity of order crossover

not associable with any simple fitness landscape defined as a height function on a metric space in a simple way. This is bad news for non-geometric crossovers because the alternative to a simple fitness landscape with a simple geometric interpretation is a complex topological landscape with hardly any interpretation for what is really going on.

This leads us to a third very important practical consequence. Just knowing that a recombination operator is geometric or non-geometric cannot tell us anything about its performance. The no free lunch theorem rules. However, as a rule-of-thumb we know that when the fitness landscape associated with a geometric crossover is smooth, the geometric crossover associated with it is likely to perform well. This is fundamental for crossover design because the designer studying the objective function can identify a metric for the problem at hand that gives rise to a smooth fitness landscape and then he/she can pick the geometric crossover associated with this metric. This is a good way to embed problem knowledge in the search. However, since this strategy is inherently linked to the existence of a distance function associated with a recombination operator, non-geometric crossovers cannot make use of it.

5.1   Possibility of a General Theory of Evolutionary Algorithms

The forth and last consequence of the mere existence of some non-geometric operators is that this implies the existence of two separate classes of operators. We state this in the following as a theorem.

Theorem 7 (*Existence of non-geometric crossover*). *The class of non-geometric crossover is not empty. Hence the space of recombination operator is split into two proper classes: geometric and non-geometric crossover.*

This is an important step when developing a theory of geometric crossover because it allows to meaningfully talk about geometric crossover in general without the need to specify the distance associated with it. This in turn has a critical impact on the possibility of a general theory of geometric crossover and of a programme of unification of evolutionary algorithms.

The main danger of a general theory is being too shallow: would such a general theory be able to tell us anything meaningful or only trivialities encompassing all operators could be derived? A theory of all operators is an empty theory because the performance of an EA derives from how its way of searching the search space is matched with some properties of the fitness landscape. Without restricting the class of operators to a proper subset of all possible operators, there is no common behavior, hence there is no common condition on the fitness landscape to be found to guarantee better than random search performance. This is just another way of stating the NFL theorem. So a theory of all operators is necessarily a theory of random search in disguise. However, since the definition of geometric crossover does not encompass all operators, it is not futile to pursue a general theory of geometric crossover.

In previous work we have found that many recombinations used in every-day practice are geometric. Without being able to prove the existence of some non-geometric crossovers there are two alternative explanations for this happening: (a) the geometric crossover definition is a tautology and the theory built on it a theory of everything hence an empty theory or (b) if there are non-geometric crossovers, this is hardly a coincidence and the class of geometric crossover indeed captured a deep aspect of the class of "real-word" recombinations.

Theorem 7 is therefore foundational because it implies that the true explanation is (b). Therefore, a general theory of geometric crossover makes sense because it is not a theory of random search in disguise and the program of geometric unification of evolutionary algorithms makes sense because it is not a mere tautology.

6   Conclusions and Future Work

In this paper we have shown that the abstract definition of geometric crossover induces two *non-empty* representation-independent classes of recombination operators: geometric crossovers and non-geometric crossovers. This is a fundamental result that put a programme of unification of evolutionary algorithms and a general representation-independent theory of recombination operators on a firm ground.

Because of the peculiarity of the definition of geometric crossover, proving non-geometricity of a recombination operator, hence the existence of the non-geometric crossover class, is a task that at first looks impossible. This is because one needs to show that the recombination considered is not geometric under *any distance*. However taking advantage of the different possible ways of looking at the definition of geometric crossover we have been able to develop some theoretical tools to prove non-geometricity in a straightforward way. We have then used these tools to prove the non-geometricity of three well-known operators for real vectors, permutations, and syntactic trees representations.

In future work, we will start constructing a general theory of evolutionary algorithms based on the abstract interpretation of the definition of geometric crossover. So this theory will be able to describe the generic behavior of all evolutionary algorithms equipped with a generic geometric crossover. We anticipate that this is a form of convex search. The next step will be to understand for what general class of fitness landscape this way of searching delivers good performance.

References

1. Bäck, T., Hammel, U., Schwefel, H.: Evolutionary computation: Comments on the history and current state. IEEE Transactions on Evolutionary Computation 1(1), 3–17 (1997)
2. Bäck, T., Fogel, D.B., Michalewicz, T.: Evolutionary Computation 1: Basic Algorithms and Operators. Institute of Physics Publishing (2000)
3. Blumental, L., Menger, K.: Studies in geometry. Freeman and Company, San Francisco (1970)

4. Deza, M., Laurent, M.: Geometry of cuts and metrics. Springer, Heidelberg (1991)
5. Jones, T.: Evolutionary Algorithms, Fitness Landscapes and Search. PhD dissertation, University of New Mexico (1995)
6. Kim, Y.H., Yoon, Y., Moraglio, A., Moon, B.R.: Geometric crossover for multiway graph partitioning. In: Proceedings of the Genetic and Evolutionary Computation Conference, pp. 1217–1224 (2006)
7. Lucas, S., Moraglio, A., Li, H., Poli, R., Zhang, Q.: Alignment Crossover for Finite State Machines. (in preparation)
8. Luke, S., Spector, L.: A Revised Comparison of Crossover and Mutation in Genetic Programming. In: Proceedings of Genetic Programming Conference (1998)
9. Moraglio, A., Poli, R.: Topological interpretation of crossover. In: Proceedings of the Genetic and Evolutionary Computation Conference, pp. 1377–1388 (2004)
10. Moraglio, A., Poli, R.: Topological crossover for the permutation representation. In: Workshop on theory of representations at Genetic and Evolutionary Computation Conference (2005)
11. Moraglio, A., Poli, R.: Geometric Crossover for the Permutation Representation. In: Technical Report CSM-429, University of Essex (2005)
12. Moraglio, A., Poli, R.: Geometric Landscape of Homologous Crossover for Syntactic Trees. In: Proceedings of IEEE Congress of Evolutionary Computation, pp. 427–434 (2005)
13. Moraglio, A., Poli, R.: Product Geometric Crossover. In: Proceedings of Parallel Problem Solving from Nature, pp. 1018–1027 (2006)
14. Moraglio, A., Poli, R.: Geometric Crossover for Sets, Multisets and Partitions. In: Proceedings of Parallel Problem Solving from Nature, pp. 1038–1047 (2006)
15. Moraglio, A., Poli, R., Seehuus, R.: Geometric Crossover for Biological Sequences. In: Proceedings of Genetic Programming Conference, pp. 121–132 (2006)
16. Moraglio, A., Togelius, J., Lucas, S.: Product Geometric Crossover for the Sudoku Puzzle. In: Proceedings of IEEE Congress of Evolutionary Computation, pp. 470–476 (2006)
17. Pardalos, P.M., Resend, M.G.C.: Handbook of Applied Optimization. Oxford University Press, Oxford, UK (2002)
18. Radcliffe, N.: Equivalence Class Analysis of Genetic Algorithms. Journal of Complex Systems 5, 183–205 (1991)
19. Reidys, C.M., Stadler, P.F.: Combinatorial landscapes. SIAM Review 44, 3–54 (2002)
20. Seehuus, R., Moraglio, A.: Geometric Crossover for Protein Motif Discovery. In: Workshop on theory of representations at Genetic and Evolutionary Computation Conference (2006)
21. Syswerda, G.: Schedule Optimization Using Genetic Algorithms. In: Davis (ed.) Handbook of Genetic Algorithms, Van Nostrand Reinhold, New York (1990)
22. Van de Vel, M.L.J.: Theory of Convex Structures. North-Holland (1993)
23. Wright, S.: The roles of mutation, inbreeding, crossbreeding and selection in evolution. In: Jones, D.F. (ed.) Proceedings of the Sixth International Congress on Genetics, vol. 1, pp. 356–366 (1932)

Just What Are Building Blocks?

Christopher R. Stephens[1] and Jorge Cervantes[2]

[1] Instituto de Ciencias Nucleares, Universidad Nacional Autónoma de México
Circuito Exterior, A. Postal 70-543, México D.F. 04510
[2] IIMAS, Universidad Nacional Autónoma de México
Circuito Exterior, A. Postal 70-543, México D.F. 04510

Abstract. Using an exact coarse-grained formulation of the dynamics of a GA we investigate in the context of a tunable family of "modular" fitness landscapes under what circumstances one would expect recombination to be "useful". We show that this depends not only on the fitness landscape and the state of the population but also on the particular crossover mask under consideration. We conclude that rather than ask when recombination is useful or not one needs to ask - what crossover masks are useful. We show that the answer to this is when the natural "building blocks" of the landscape are compatible with the "building blocks" defined by the crossover mask.

1 Introduction

Recombination has for many been the key operator which distinguishes many Evolutionary Algorithms (EAs), such as GAs, from other classes of heuristic, the idea being that recombination takes fit "partial" solutions from one section of the population and recombines them with fit "partial" solutions from another section to form more "complete" solutions. This thinking has been encapsulated in Holland's Schema theorem (HST) [1] and the associated Building Block Hypothesis (BBH) [2], where a "partial" solution is formulated in terms of the concept of a schema, the conclusion being that a GA finds optimal, or near optimal, strings by recombining fit, short low-order schemata, the short being deduced from analysing only the destructive effects of crossover.

This intuitive framework as to how recombinative EAs work has had two chief drawbacks: firstly, that exact microscopic models [3] seemed to give no evidence at all that schemata or "building blocks" naturally emerged as a preferred description of the dynamics. On the contrary, such models have been used to argue against the utility of these concepts; and secondly, that attempts to design fitness landscapes where recombinative EAs manifestly performed better than non-recombinative ones, have had, at best, ambiguous results. The most well known example of this are the "Royal Road" functions [4], [5], where "building blocks" were deliberately built into the construction of the landscape. More recently in [6] real-valued analogs of the Royal Road function were used as examples of landscapes where recombination was provably better than mutation. Other contrived examples have followed, such as the HIFF function [7]. Of course, this begs the

C.R. Stephens et al. (Eds.): FOGA 2007, LNCS 4436, pp. 15–34, 2007.

question of the utility of recombination if it is so difficult to find landscapes that favour it. One may also ask why is it then so ubiquitous in nature?

Although, exact, string-based models gave no hint of when or where recombination may be useful, later "coarse-grained" [8,9] models showed how in recombinative EAs the concept of a schema clearly emerged, schemata being the natural effective degrees of freedom for describing recombination. What is more, the very structure of the equations showed rigorously and unambiguously how, indeed, recombinative EAs build optimal solutions from Building Blocks (BBs) which in their turn could be constructed from their BBs etc., the hierarchy terminating at the most coarse-grained BBs - one-schemata. Furthermore, the set of these BBs formed an alternative coordinate basis - the Building Block Basis (BBB) - for the description of the dynamics. Although the very structure of the equations showed clearly how recombinative EAs "worked", in a manner akin to the intuition if not the mathematics behind the BBH, what the structure did not show is under what conditions recombination was useful and what BBs were favoured. Although some early empirical work showed [10] how the BBH was not generally correct in its conclusion that short blocks are preferred, not much more has been done beyond qualitative statements about natural metrics associated with the dynamical equations of recombinative EAs. This is a pity, as much insight can be gained from the exact coarse-grained equations. So, the purpose of this paper is to give insight as to when one would expect recombination to be useful by examining the equations for a recombinative GA in a set of model landscapes and using a set of model metrics to measure the utility of recombination.

Although we have extensive results for many different forms of crossover and many different fitness landscapes, due to space limitations here we will present results only for one-point crossover and a family of "modular" fitness landscapes.

2   A Short History of "Building Blocks"

HST has been one of the pillars on which much of the theory of GAs has been based. The notion behind a schema, a marginal in the terminology more familiar in population genetics, is that, intuitively, as a definite subset of loci of a given string, it represents a "partial" solution, the potential full solution being represented by an entire string. For a given string with alleles of cardinality k, there are $(k+1)^N$ different schemata, a given string being a member of 2^N of them.

Mathematically, in one common formulation, HST states that: for a GA with one-point crossover, proportional selection and point mutation; and for an arbitrary schema I, of length ℓ and order N_o, I being a multi-index $I = \{i_1, i_2, \ldots, i_{N_o}\}$,

$$\langle P_I(t+1)\rangle \geq \left(\left(1 - p_c \frac{(\ell-1)}{N-1}\right)\right)(1-p)^{N_o}\frac{f(I)}{\bar{f}(t)}P_I(t) \tag{1}$$

where $P_I(t)$ is the frequency of the schema I at generation t. The parameters p_c and p are the recombination and mutation rates respectively, while $f(I)$ is the schema fitness and $\bar{f}(t)$ the average population fitness. This version of the Schema theorem is an inequality as it does not take into account the effects of schema creation.

At least with respect to recombination, the canonical interpretation of equation (1) is that the longer the schema the less chance it has of being propagated into the next generation due to schema disruption. Similarly, the effect of mutation has been interpreted so as to favour low-order schemata. Hence, as clearly the higher the fitness of a schema the more likely it is to propagate, the above has been formalized in terms of the Building Block Hypothesis

- A GA works by recombining short, low-order, highly fit schemata into even fitter higher-order schemata

A "building block" in this traditional sense is taken to be a fit, low-order, short schemata, such that when juxtaposed via recombination with other like schemata helps find fitter solutions. As indicated above, this characterization of the particular schemata that should be termed "building blocks" comes from a particular interpretation of HST – schema disruption leading to the idea of short and preservation under mutation leading to the idea of low order. However, HST is incapable of answering how the juxtaposition of these building blocks comes about, as it does not account for this part of the process. This is a major weakness when considering HST and the BBH as the basis on which to explain how a GA works. Although, generally accepted in many quarters there have been other alternative interpretations [11,12] proffered as to how a GA "works".

Another subtlety concerns the fitness $f(I)$ of a schema or building block. In (1) this fitness is population dependent, being given by $f(I) = \sum_{J \in I} f(J)P_J(t)/\sum_{J \in I} P_J(t)$ where the sum is over all strings J that are elements of the schema I. This dynamic schema fitness must be contrasted [13] with the static schema fitness $f_s(I)$, which can be found from the dynamic one by setting the population to be random, wherein the $P_J(t)$ cancel. It also needs to be emphasized that the BBH does not provide a mathematically rigorous characterization of building blocks. For instance, although we know how many schemata there are, we do not know how many building blocks there are, as we do not have a precise definition with which to establish a subset. Just how short does a building block have to be to be considered such. There are clearly however $(k+1)^N$ potential building blocks.

A corollary of (1) has been taken to be that it is advantageous to put as close together as possible bits that must "cooperate". This is stated so that one requires to increase the "linkage" between cooperating genes so that there is less probability of splitting them up during recombination. This notion of linkage is one that is shared by population genetics and is intimately associated with the act of recombination. However, the term "linkage" has also been widely used in the EA community to denote epistatic dependence between genes [14]. In other words, if the fitness contribution associated with a given locus depends on another then they are said to be "linked". Linkage in this sense is associated with the relation between a fitness value and a representation. For instance, a separable fitness function can be decomposed into a set of linkage units or blocks wherein there is epistasis between loci within a block but not between blocks. It is important to remember, however, that a block in this sense is not necessarily the same as a building block as defined by HST and the BBH. Nor, as we will see in the next section, is it the same as a Building Block which is an element of the BBB To distinguish them we will refer to a linkage group with mutual epistasis as a landscape block.

It is then natural to think that a good search procedure should respect a given landscape block once its optimal configuration has been found. Hence, an understanding of the linkage patterns associated with a given representation can help in finding good search algorithms for the problem. Such "linkage learning" has spawned many different algorithms [14]. Much of this work is associated with the question of how genes should be best distributed on a string. This is a question of representation.

Which representation is best however, depends on what genetic operators are to be used for the search. The BBH, based on HST, would hint that it is better to use a representation such that linked genes (in the epistatic sense) are placed close to one another on the string. This would mean that the associated land-scape blocks are as short as possible. What is at issue from the point of view of the design of efficient search algorithms is a good compatible choice of repre-sentation and genetic operators. From the point of view of selecto-recombinative algorithms this seems to be hinting at the fact that landscape blocks and build-ing blocks should be as compatible as possible. What can we play with though to enhance this compatibility? To some extent our freedom is limited.

3   What Do Exact Models Tell Us?

In the previous section we talked about the notion of building block in the context of HST. As emphasized, this theorem does not tell us about schema reconstruction. For that, we need an exact model. The model we will consider will be the canonical GA with selection, mutation and homologous crossover, though here we will concentrate largely on selection and recombination. Mathematically, in the string basis, the evolution equation familiar from population genetics or GAs [3] that relates the expected value of the genotype frequency at generation $t + 1$, $\langle P_I(t + 1) \rangle$, to the actual genotype frequencies at generation t is

$$\langle P_I(t + 1) \rangle = \sum_J M_I{}^J \left((1 - p_c) P_J'(t) \right.$$

$$\left. + p_c \sum_m p_c(m) \sum_{K,L} \lambda_J{}^{KL}(m) P_K'(t) P_L'(t) \right) \qquad (2)$$

where P_I' is the probability to select I. $M_I{}^J$ is the probability to mutate genotype I to genotype J and $\lambda_I{}^{JK}(m)$ is the conditional probability that the offspring I is formed given the parents J and K and a recombination mask m, $p_c(m)$ being the conditional probability that the mask m is applied and p_c the probability that crossover is applied in the first place. $\lambda_I{}^{JK}(m) = 0$, 1 as either the offspring is formed or it isn't. In the infinite population limit, $\langle P_I(t) \rangle \to P_I(t)$ and equation (2) is a deterministic equation for $P_I(t)$.

The first term in (2) arises from mutation of a clone of the string J, while the second term represents the mutations of string J, where now J arises from all the ways in which it may be constructed from other strings via recombination. The explicit form of P'_I depends on the type of selection used. For proportional selection, for instance, which we will use here, $P'_I(t) = (f(I)/\bar{f}(t))P_I(t)$.

If one wishes to "explain" how GAs work, in one sense one need look no further than (2), given that it is an exact equation. However, explain also has a connotation of understand. It is not sufficient to just numerically integrate it, even if that were possible for larger values of N. We can however try to use it to see to what extent the historical explanation of how GAs work in section (2) is valid. If the intuitive explanation associated with HST and the BBH are valid, then they should be compatible with the rigorous theory described by (2). In fact, (2) and its consequences have been used as a stick with which to beat proponents of HST and the BBH as neither is readily visible in it. Here, string construction is interpreted as the pure recombination of other strings. There are no building blocks manifest here. What is more, if one wishes to solve these equations one must consider that the equation for a single string potentially depends on all the rest.

However, (2) is written in terms of strings. What happens if one considers schemata? Might things be simpler? Equation (2) can be coarse grained to yield an equation for an arbitrary schema, I, that has an identical functional form. For instance, for one-point crossover

$$\langle P_I(t+1) \rangle = \sum_J M_I{}^J \left(\left(1 - \frac{p_c(\ell-1)}{(N-1)} \right) P'_J(t) \right.$$
$$\left. + \frac{p_c}{N-1} \sum_k \sum_{K,L} \lambda_J{}^{KL}(k) P'_K(t) P'_L(t) \right) \qquad (3)$$

where the sum over k is over those crossover points within the defining length of the schema J. The interpretation of the resulting equation is that the first term arises from mutation of a clone of the schema J, while the second term represents the mutations of J, where now schema J arises from all the ways in which it may be constructed via recombination from other schemata of the *same* schema partition. By neglecting for the moment the schema reconstruction term and mutations from other schemata J one obtains (1). Thus, one may recover HST. The question remains however, are building blocks really used and if so how are they recombined? Any answer has to lie in the construction terms of (2) and (3). However, as in the case of strings these just say that schemata are mixed together by recombination to form other schemata. If one wants to understand how a particular schema arises one must consider all other schemata on the partition. Furthermore, how would one naturally obtain information about strings once one has passed to a description such as this in terms of schemata? Of course, one can reasonably surmise that the strings/schemata that are mixed should be fit, as they appear in factors P'_K and P'_L, but what about short or low order?

The answer to these questions can be more easily attacked by passing to a different representation of (2). This is found by noting that the string reconstruction term itself can be naturally written in terms of schemata. This can be achieved in two different but equivalent ways, one by a direct coarse graining of the construction term [8], or secondly, via a coordinate transformation [15] implemented by a coordinate transformation matrix $\Lambda_1 \equiv \begin{pmatrix} 1 & 1 \\ 0 & 1 \end{pmatrix}^{\otimes N}$ where $\otimes N$ signifies the N-fold tensor product of the matrix[1]. In the absence of mutation (for clarity) one finds

$$\langle P_I(t+1)\rangle = P'_I(t) - p_c \sum_m p_c(m)\Delta_I(m,t) \tag{4}$$

where

$$\Delta_I(m,t) = (P'_I(t) - P'_{I_m}(t)P'_{I_{\bar{m}}}(t)) \tag{5}$$

is the selection-weighted linkage disequilibrium coefficient [8]. In this representation I_m and $I_{\bar{m}}$ are conjugate schemata, i.e., $I_{\bar{m}}$ is the bit complement of I_m in the string I. For example, $*1$ is the conjugate of $1*$ in the string 11. $P'_{I_m}(t)$ is the probability to select the schema I_m and $P'_{I_{\bar{m}}}(t)$ that of the conjugate schema $I_{\bar{m}}$. For example, for $N = 3$, if the optimal string is $I = 111$ and the mask is 011, $0/1$ signifying take the corresponding bit from the first/second "parent", then the schema defined by the mask is $I_m = 1 * *$ (just replace mask 0s by *s in the schema). As $\bar{m} = 100$, then the conjugate schema is $I_{\bar{m}} = *11$, such that $I = I_m \cup I_{\bar{m}}$. Hence, $\Delta_{111}(011,t) = P'_{111}(t) - P'_{1**}(t)P'_{*11}(t)$.

The most noteworthy thing about (4) compared to (2) is that string reconstruction is now seen to proceed via the recombination of schemata, string I being formed by recombining the schemata I_m and $I_{\bar{m}}$, which indeed in a very real sense are now the "building blocks" of the string I. However, in order to understand the dynamics of I we have to know in turn the dynamics of I_m and $I_{\bar{m}}$. The equations for these have the same functional form as (4). The reconstruction term for the schema I_m, for instance, now involves two other conjugate schemata $I_{mm'}$ and $I_{m\bar{m}'}$. For example, for $I = 11*$ and $m' = 010$ or 011 then $I_{m'} = 1 * *$ and $I_{\bar{m}'} = *1*$. Thus, we see that the building blocks of I in turn have their own building blocks, which are of lower order than I_m, which in its turn is of lower order than I.

Inherent in (4) then is an important element of the BBH - that a GA builds solutions by recombining lower order partial solutions. What we cannot naively conclude is that these partial solutions are fit, short or low-order. However, what the equation does show is that for a selecto-recombinative GA, if a string is not present

---

[1] In this basis the sums over K and L can be dispensed with, as the three-index object $\lambda_J{}^{KL}(m)$ is skew-diagonal on the indices K and L, coupling only the two schemata, I_m and $I_{\bar{m}}$, that contribute to the string I. Moreover, for a given mask, there is one and only one corresponding schema combination that yields the string I.

in the population, then the only way to find it is by recombining its component schemata and, further, that the *only* way that these component schemata can be found, if they are not already present, is by recombining their own lower-order component schemata. For a given string, there is a unique set of 2^N definite schemata that can be potentially used by recombination to reconstruct it. What is more these schemata form a basis set with which the dynamics of any other string may be deduced. For these reasons we denote these schemata Building Block schemata.[2] Of course, there is a corresponding BBB for any given string. They all form equivalent bases. Naturally, a very important one is that associated with the optimal string, for which the corresponding BBB gives the unique set of schemata that potentially can be recombined to obtain it. If our interest is in the optimal string, then (4) tells us that instead of the $(k+1)^N$ potential building blocks (remember no capitals!) only 2^N can possibly contribute to the dynamics of the optimal string. Furthermore, as we shall see, this set of interest may be further reduced in size after considering the particular recombination distribution used.

The importance of $\Delta_I(m, t)$ is that it offers a complete description of the utility of recombination, mask by mask and generation by generation. For instance, if $\Delta_I(m, t) < 0$, then recombination using the mask m produces more strings of type I in the next generation than selection alone. Similarly, if $\Delta_I(m, t) > 0$ the contrary is true. Clearly Δ_I depends on the fitness landscape and the actual state of the population. However, it also depends explicitly on the mask, which means that in evaluating the effects of recombination it is necessary to consider it mask by mask, as potentially one mask may be advantageous while another is disadvantageous.

3.1   The Landscapes

As the results of recombination and selection are sensitive to the fitness landscape chosen one has to focus on a subset of possible landscapes that illustrate important relationships between the two. Here, we will focus on a class of "modular" landscapes of the following type:

$$f_I = \sum_{\alpha=1}^{M} f_{I_\alpha}$$

where the string is divided up into M consecutive, contiguous landscape blocks, f_{I_α} being the fitness of block α, $\alpha \in [1, M]$. It remains to specify the fitness distribution within a block. We will take the landscape to be uniform, in that each block has the same landscape and will principally consider the block-landscape to be needle-in-a-haystack (NIAH)[3]. Thus,

$$f_{I_\alpha} = f_1 \quad I_\alpha = \text{needle} \tag{6}$$
$$= f_0 \quad I_\alpha = \text{hay} \tag{7}$$

---

[2] We capitalize "building block" now as in this case they have a mathematically rigorous foundation, forming an alternative coordinate basis - the Building Block basis (BBB) [15] - for the description of the dynamics.

[3] We have tried other landscapes, such as a deceptive landscape, with similar results.

This class of landscapes is clearly "tunable" in its modularity through M: $M = 1$ being just the standard NIAH landscape, while for $M = N$ it leads to a unitation landscape, being counting ones for $f_1 = 1$, $f_0 = 0$. Generally we will take for each block the optimal block configuration to be all 1s and the needle to have fitness 2, and the hay fitness 1. If $f_0 = 0$ for $M \neq N$ it is the "Royal Road" function. If the landscape within each block is the N/M-bit deceptive landscape, then the whole landscape becomes the concatenated-trap function with m traps.

Although the modularity we have for simplicity chosen here is "absolute", in that there are no epistatic interactions between blocks, our results will in general be qualitatively valid for landscapes where there are "weak" interactions between blocks.

3.2   The Metrics

We now turn to the question of how performance will be measured. The metrics we will use are:

1. $\Delta_I(m, t) = P'_I(t) - P'_{I_m}(t) P'_{I_{\bar{m}}}(t)$, which, as explained in section 3, measures how recombination contributes to the frequency change of a given string, schema or BB I from generation t to generation $t + 1$.
2. $\mathcal{P}_I(O_1, O_2, t) = P_I^{O_1}(t) / P_I^{O_2}(t)$ measures the relative effect of a set of operators O_1 with respect to another set O_2 in the production of a given string, schema or BB I from generation t to generation $t + 1$. By normalizing with respect to the operator set O_2 we can isolate the effects of the operators that are different between the two sets. For instance, if O_1 represents selection and recombination, while O_2 represents selection only, then $\mathcal{P}(O_1, O_2, t)$ represents the proportion of objects of type I produced by recombination in the presence of selection relative to the proportion produced by selection alone.
3. Distance of the population at time t from the Geiringer manifold $d_G = (\sum_I (P_I(t) - P_G(t))^2)^{1/2}$, where $P_G(t) = \prod_{i=1}^{N} P_{I_i}(t)$ gives the point on the Geiringer manifold associated with the current population.
4. Distance of the population at time t from the center of the simplex, $d_C = (\sum_I (P_I(t) - P_C)^2)^{1/2}$, where $P_C = 1/2^N$ is the frequency of each string in a random population.
5. Distance of the population at time t from an ordered population at the optimal vertex of the simplex, $d_O = (\sum_I (P_I(t) - P_O)^2)^{1/2}$, where $P_O = \delta_I^{\text{opt}}$ is 1 when the entire population is at the optimal vertex opt and zero otherwise.

3.3   What's the Point?

It should be emphasized that (4) gives an exact description of the dynamics in the infinite population limit. However, one is confronted by the question - so what? What can these equations tell us? There are answers to this question at different levels of abstraction. First of all, the equations tell us indeed how, in principle, a recombinative GA builds up "solutions" (optimal or near-optimal strings) by recombining "partial solutions" (BBs) of the solution. Furthermore, the BBs of that solution could also have been produced by their own BBs etc. until one arrives at the "Adam and Eve" schemata - the one-schemata, the most primitive members of the BB hierarchy. In fact, if the solution isn't already present, then, in the absence of mutation, construction of a "good" solution by joining together its BBs is the *only* way of getting it. Furthermore, if there do not exist pairs of conjugate BBs in the population with which I can be obtained, then the *only* way to obtain these BBs is to recombine their BBs.[4]

This can be simply illustrated with a concrete example. Consider for $N = 3$ and one-point crossover, the evolution of a "good" solution, which we arbitrarily take to be 111, then

$$P_{111}(t+1) = (1-p_c)P'_{111}(t) + \frac{p_c}{2}(P'_{1**}(t)P'_{*11}(t) + P'_{11*}(t)P'_{**1}(t)) \quad (8)$$

The right hand side of this equation depends on $P_{1**}(t)$, $P_{*11}(t)$, $P_{11*}(t)$ and $P_{**1}(t)$. The corresponding equations for them, and the other elements of the BBB are

$$P_{11*}(t+1) = (1 - \frac{p_c}{2})P'_{11*}(t) + \frac{p_c}{2}P'_{1**}(t)P'_{*1*}(t) \quad (9)$$

$$P_{1*1}(t+1) = (1-p_c)P'_{1*1}(t) + p_cP'_{1**}(t)P'_{**1}(t) \quad (10)$$

$$P_{*11}(t+1) = (1 - \frac{p_c}{2})P'_{*11}(t) + \frac{p_c}{2}P'_{*1*}(t)P'_{**1}(t) \quad (11)$$

$$P_{1**}(t+1) = P'_{1**}(t) \quad (12)$$

$$P_{*1*}(t+1) = P'_{*1*}(t) \quad (13)$$

$$P_{**1}(t+1) = P'_{**1}(t) \quad (14)$$

$$P_{***}(t+1) = P'_{***}(t) = 1 \quad (15)$$

Thus, we can clearly see how a solution can be built from its BBs, which in their turn can be constructed from their BBs. The most coarse-grained BBs, the one-schemata, satisfy homogeneous (they don't have BBs) equations that for a given generation are explicitly independent of p_c. Note that the BBB also tells us that a complete description of how to obtain a good solution depends only on the 8 BBs of the solution not on any of the other 19 ((27-8)) possible 3-bit schemata that are not BBs of 111. Thus, for example, the schema $0 * 1$ cannot enter into the dynamics of the solution 111, or of any of its BBs.

It is worth noting that there are also BBs for selection only dynamics. The chief difference compared to recombinative dynamics is that they cannot be joined together. However, considering the selection only case allows us to deduce which BBs are preferred in the selection process before being put together. Thus, for example, for the modular landscapes defined in section 3.1, we can determine which optimal BBs are preferred, i.e., are selected with higher probability. To understand the bias of selection for BBs we consider a random population and calculate the fitnesses of the different optimal BBs in each

---

[4] These statements are just as true for a finite population as an infinite population, the only difference there being that in a finite population the actual number of times a BB is sampled could be quite different to the expected number.

landscape. For illustration, for $N = 4$, for one landscape block (i.e., NIAH):
$f_{1111}(= 2) > (f_{111*} = f_{1*11} = f_{11*1} = f_{*111}(= 3/2)) > (f_{11**} = f_{1*1*} = f_{1**1} = f_{*11*} = f_{**11} = f_{*1*1}(= 5/4)) > (f_{1***} = f_{*1**} = f_{**1*} = f_{***1} (= 9/8)) > f_{****}(= 17/16)$. Thus, higher order optimal BBs will be selected preferentially. For four landscape blocks (i.e., counting ones) the analogous hierarchy is: $f_{1111}(= 4) > f_{111*}(= 7/2) > f_{11**}(= 3) > f_{1***}(= 5/2) > f_{****}(= 2)$, where, as in the one block case, each BB of a fixed order has the same fitness, e.g. $f_{11**} = f_{1*1*}$. So, for these landscapes there is no dependence on the arrangement of the optimal alleles, only on their number. Consider now though the landscape with 2 NIAH blocks. In this case, as with the one-block and four block landscapes $f_{1111} > f_{111*} > f_{11**} > f_{1***} > f_{****}$, but now, in distinction, $f_{11**}(= 13/4) > f_{1*1*}(= 3)$. The reason for this is, of course, obvious, the BB $11 * *$ respects the block structure of the fitness landscape, i.e. the BB $11 * *$ is *also* a complete landscape block. On the other hand, $1 * 1*$ is not a complete landscape block but rather corresponds to two suboptimal blocks. Thus, the modular nature of the landscapes we are considering define corresponding preferred optimal BBs.

Having seen that in the dynamics of selection and recombination the only way to get a good solution is to either start with it or construct it from its BBs we can now try to answer the more specific questions of: i) under what circumstances recombination is useful? ii) what schemata or BBs are useful?

Point i) has a first obvious answer - when the solution or desired BB is not already present. However, does this mean that if it is present then recombination is not useful? As can be clearly seen from (4) the utility of recombination, i.e., the sign of Δ_I, in a given generation and for a given string or BB, depends both on the fitness landscape and the current population. If $P_I(t) = 0$, then $\Delta_I(m,t) \geq 0 \ \forall m$, and recombination is a positive or neutral effect. However there exists a critical value of P_I, P_I^c, given by

$$P_I^c = \frac{f_{I_m} P_{I_m} f_{I_{\bar{m}}} P_{I_{\bar{m}}}}{f_I \bar{f}} \tag{16}$$

such that for $P_I \geq P_I^c$, $\Delta_I \geq 0$ and hence recombination and selection produce less copies of the good solution than selection alone. The question then is: is this existence of a critical frequency a landscape independent effect? To investigate this we will "neutralize" the population effect by putting in a random population, i.e., $P_I(t) = 1/2^N$. Then,

$$\Delta_I(m,t) = \frac{(f_I \bar{f}(t) - f_{I_m}(t) f_{I_{\bar{m}}}(t))}{2^N \bar{f}^2(t)} \tag{17}$$

Two simple landscapes are the NIAH and counting ones landscapes. Considering first NIAH, for $N = 4$ and one-point crossover, where the optimum is 1111, then $f_{1111} = 2$ and $\bar{f} = 17/16$. For $m = 0001$, $f_{111*} = 3/2$ and $f_{***1} = 9/8$, hence, $\Delta_{1111}(0001) = 7/17^2 > 0$ and so we conclude that for a random (infinite) population in a NIAH landscape recombination is disadvantageous. Δ_{1111} can similarly be shown to be greater than zero for any mask. On the other hand for counting ones: $f_{1111} = 8$ and $\bar{f} = 6$. For $m = 0001$, $f_{111*} = 15/2$ and $f_{***1} = 13/2$. Thus, $\Delta_{1111}(0001) = -3/2304 < 0$, therefore we see that for this mask recombination is advantageous, in that it leads to more optimal strings than pure selection. Once again, it is simple to show, for any N and any m, that Δ_{1111} is always negative. This also holds for any BB of the optimal string. As a final example, let's consider $N = 4$, $M = 2$. In this case, $f_{1111} = 4$ and $\bar{f} = 5/2$. For $m = 0001$, $f_{111*} = 7/2$ and $f_{***1} = 11/4$ from which we deduce that $\Delta_{1111}(0001) = 9/1600 > 0$, and therefore that for this mask recombination is disadvantageous. On the other hand, for $m = 0011$, one finds $f_{11**} = 13/4$ and $f_{**11} = 13/4$ which yields $\Delta_{1111}(0011) = -9/1600 < 0$. Thus, we see that in general the utility of recombination depends not only on the fitness landscape and the population but also on the specific mask used. In this case, the mask 0001 leads to a negative effect while the mask 0011 leads to a positive one. The reason for this is that the mask 0001 does not respect the boundaries of the natural blocks as defined by the fitness landscape, which in this case are the first two loci as one block and the last two as another. The mask 0011 on the other hand does respect the landscape blocks and therefore is less destructive.

4   Results

4.1   Benchmarks

We begin here by considering two "benchmarks": what happens in the presence of selection only and what happens in the presence of recombination and/or mutation in the absence of selection. In this sense we are looking at the "intrinsic" biases of the individual operators without regard as to how they interact. We start off with Figure 1, seeing the effects of recombination and mutation in a flat fitness landscape. We consider the system with $p_c = 0$, 1 and $p = 0$, 0.1. Shown are the three distance measures d_G, d_C and d_O. As we know analytically that $d_G = d_C = 0$ for a random population (this corresponds to the center of the simplex which is also a point on the Geiringer manifold) we consider an initial population that is non-random, choosing $P_{0000}(0) = 0.2$, $P_{1111}(0) = 0.8$. As the initial population is not on the Geiringer manifold we can see that in the absence of mutation or recombination the distance to the Geiringer manifold or the center stays constant. Recombination moves the population in the direction of the Geiringer manifold exponentially quickly, $d_G \sim (0.67)^t$. In contrast, the population nears the center initially, before eventually settling down at a constant distance from it. For mutation in the absence of recombination, one can see that the population approaches the Geiringer manifold and the center exponentially quickly, with rates $(0.64)^t$ for the Geiringer manifold, and $(0.80)^t$ for the center, showing that with mutation alone the system approaches the Geiringer manifold even quicker than it approaches the center. At least with this quite high mutation rate then, we see that mutation takes the population more quickly towards the Geiringer manifold than the center. Finally, with recombination and mutation together we see that the population approaches the Geiringer manifold as $(0.42)^t$, i.e. the two acting together give an uncorrelated population much quicker

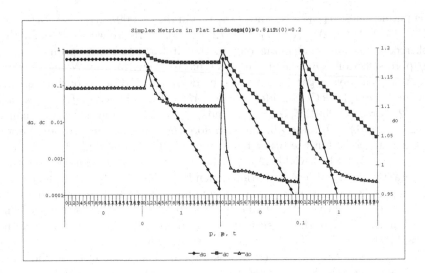

Fig. 1. Time evolution of the simplex metrics for a flat landscape and inhomogeneous initial population. Two recombination rates - 0 and 1 - and two mutation rates - 0 and 0.1 are shown.

than either separately. Also, we can see that recombination does not move the system in the direction of the center, so mutation can take you to the Geiringer manifold but recombination can't take you to the center.

In Figure 2 we see how different Δ_I evolve under the same conditions as the distance measures of Figure 1. We can see that in the absence of recombination and mutation correlations in the population are preserved. In the presence of recombination however, the Δ_I tend to zero, as in this case the system approaches the Geiringer manifold, where by definition $\Delta_I = 0$. We can see that all the correlations eventually decay exponentially rapidly with the same fixed exponent. The correlations for 11*, 111* and 1111 all decay the same asymptotically because they are dominated by the decay rate of 11 * *. On the other hand the Δ for 1 * 1* decays more rapidly because there are more cutting points for this schema when compared to the BB 11 * *. Interestingly, for 11 * 1 its decay rate depends on the mask used. If the mask is 01 * 1, i.e., the first two bits are cut then things decay less rapidly than with the mask 00 * 1. This is because the BB *1 * 1 decays more rapidly than the BB 11 * *. In the presence of pure mutation, once again the Δ_I go to zero, while with mutation and recombination together we get the same exponential decays as with recombination alone but now the transients decay more rapidly.

In Figure 3 we turn to the question of whether in the presence of pure selection certain BBs are preferred. We showed analytically in section 3.3 how higher order BBs were selectively preferred from a random population for NIAH and counting ones. Here we consider the evolution of different optimal BBs as a function of time, optimal meaning they are BBs of the optimal string. Graphed here are

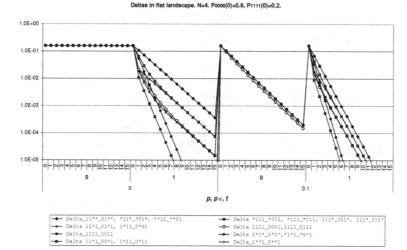

Fig. 2. Graph of Δ_I in flat landscape with inhomogeneous initial population. The notation is such that the first four symbols refer to the schemata and the second four to the mask set. The curves for the masks of a mask set are equivalent, e.g. for mask set $00*1$ the curves for 0001 and 0011 are the same.

normalized BB frequencies, the normalization being with respect to the number of strings that contribute to the BB. This normalization assures that we do not conclude that a block is preferred just because it is low order in that there are more strings that contribute to it.

The fitness landscapes chosen here are for $N = 4$ with one block (NIAH), two blocks and 4 blocks (counting ones). As can be seen in the graph for all three landscapes and in concordance with our theoretical prediction there is a preference for higher order optimal BBs that are closer to the optimal string (1111). This is because the effective fitness (which in this case of selection only dynamics is the same as the schema fitness) is greater for higher order BBs. However, in distinction with the NIAH and unitation cases, for the landscape with 2 NIAH blocks, and as predicted, the evolution of the BBs depends on the distribution of optimal alleles across the different blocks, optimal alleles that form complete optimal landscape blocks being preferred to those that are split across different landscape blocks. These results generalize to any value of M, the extreme cases of 1 or N blocks being the only ones where the distribution of optimal alleles does not influence the evolution.

4.2   Dependency on Landscape Modularity

We now turn to consider the question of how recombination and selection interact. We already predicted in section 3.3 that in the NIAH landscape once the optimal string was found, and reached a certain critical frequency, that recombination would be disadvantageous, as $\Delta_{I_{\mathrm{opt}}}(t, m) > 0$ for any m. Similarly, we

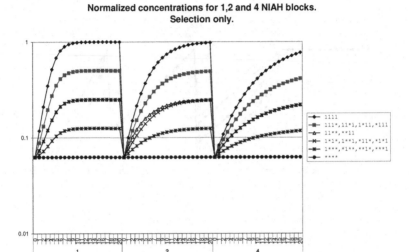

Fig. 3. Graph of normalized BB frequency for selection only dynamics for modular landscapes with 1 (left), 2 (middle) and 4 (right) landscape blocks

predicted that for counting ones recombination would *always* be beneficial. Here we will examine these predictions in terms of an integration of the equations.

In Figure 4 we plot $\mathcal{P}_I(s + r, s, t)$ as a function of time for different optimal BBs. In integrating the equations, one-point crossover and the same random population initial condition were used, irrespective of the set of operators present. As we can see, $\mathcal{P}_I(s + r, s, t) < 1$ for all optimal BBs, thus showing that recombination is disadvantageous in the presence of the optimal string. Furthermore, the relative disadvantage depends on both the order and size of the BB. Thus, the higher the order the bigger the disadvantage and the longer the length for a given order the bigger the disadvantage. Thus, among the optimal order-two schemata $P_{11**} > P_{1*1*} > P_{1**1}$. All of this is in line with the conclusions of the BBH - smaller, fit (optimal) BBs are preferred. One distinction though is that the BBH as based on HST would also consider $1 * 1*$ and $1 * *1$ as "building blocks". Although, they are elements of the BBB, for one-point crossover they cannot appear as BBs of any higher-order BB of the optimal string or the optimal string itself. Hence, they do not enter into the dynamics of the optimal string.

In Figure 5 we consider the same plot as for Figure 4 but now for a counting ones (four-block) landscape. Now we see the completely opposite effect to that of the one-NIAH block landscape, i.e., that recombination is advantageous in the presence of the optimal string and therefore is *always* advantageous in creating optimal strings. Additionally, the higher the order of the BB or schema and the longer the length for a given order the greater the advantage. Thus, in this case, for example, $P_{1**1} > P_{1*1*} > P_{11**}$. This is contrary to the BBH, the reason being that the longer the schema or BB the easier it is to construct from its BB constituents. There is another interesting effect to notice here too: that

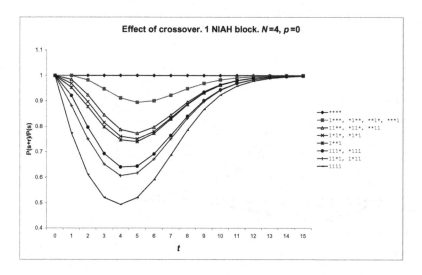

Fig. 4. Graph of $P_I(s + r)/P_I(s)$ as a function of time for $N = 4$, 1 NIAH block and one-point crossover

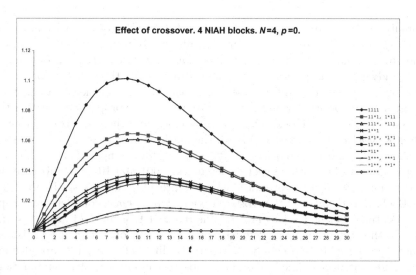

Fig. 5. Graph of $P_I(s + r)/P_I(s)$ as a function of time for $N = 4$, 4 NIAH blocks and one-point crossover

not all order-one BBs have the same evolution, BBs associated with loci at the boundary being preferred to those away from the boundary. This is not a direct effect of recombination, as p_c does not explicitly enter in the equations for the order-one BBs, but rather an indirect one, in that it enters in the schema fitness of the BBs. Thus, in this case, $f_{1***} > f_{*1**}$.

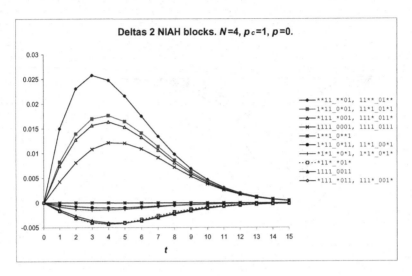

Fig. 6. Graph of $P_I(s + r)/P_I(s)$ as a function of time for $N = 4$, 2 NIAH blocks and one-point crossover. The notation is I_m, i.e. the string or schema being constructed and the corresponding mask.

Finally, we turn to the case of $N = 4$ and 2 NIAH blocks. In Figure 6 we see a graph of Δ_I as a function of time for different optimal BBs and schemata. Remembering that if Δ_I is negative recombination is disadvantageous and if positive advantageous we now see the interesting feature that whether or not recombination is useful depends precisely on which crossover mask is chosen. What distinguishes the positive from the negative curves is the property of whether the mask cuts a landscape block. Thus, for this landscape there are two order-two landscape blocks - $11**$ and $**11$. All the masks that lead to $\Delta_I > 0$ are such that they cut a landscape block of this type. Thus, for example $*111_*001$ means the optimal BB $*111$ with respect to the masks 0001 and 1001. In this case the two relevant BBs for the optimal block $*111$ are $*11*$ and $***1$. In this case the BB $*11*$, as defined by the crossover operation, is *not* a complete landscape block. On the other hand, for $*111_*011$ the relevant BBs for $*111$ are $*1**$ and $**11$. In this case, the BB as defined by the crossover operation, $**11$, is also a landscape block. It is precisely the compatibility between landscape defined blocks and crossover defined BBs that leads to the efficacy of recombination. Thus, one should not ask so much whether recombination is good or bad but rather if a certain recombination mask is good or not.

5   Finite Population Effects

In this section we will compare the infinite population model results with those using finite populations. We first check the validity of the model by comparing with a large population size as plotted in Figure 7 which is analogous to Figure 5

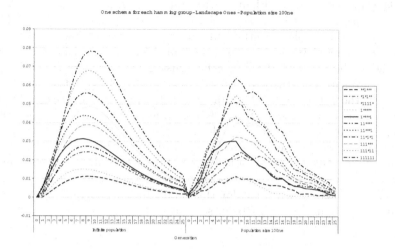

Fig. 7. Graph of $P_I(s+r)/P_I(s) - 1$ as a function of time for $N = 6$, 4 NIAH blocks (counting ones) and one-point crossover for both a large finite (left) and infinite (right) populations. Only optimal BBs are shown.

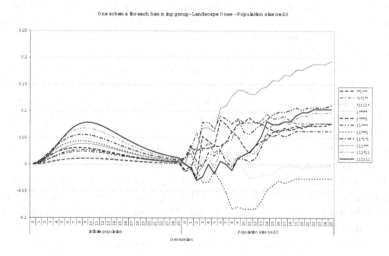

Fig. 8. Graph of $P_I(s+r)/P_I(s) - 1$ as a function of time for $N = 6$, 4 NIAH blocks (counting ones) and one-point crossover for both infinite (left) and small finite (right) populations. Only optimal BBs are shown.

for the case of counting ones. The population was set to be 100 times the total number of possible states, i.e., $100 * 2^N$. We plot the average of 5 runs starting from a random population. For this large population size, even though the number of runs is small, we can see a clear correspondence between the finite and infinite population results.

The principle feature of Figure 7 with respect to the effect of recombination is that we can clearly see the positive effect of recombination, as $P_I(s+r)/P_I(s) - 1 > 0$ over the whole evolution. In this case the use of lower order BBs to construct higher order ones is aiding the search for the optimum. What is more, the higher the order of the BB being constructed the more beneficial is the effect of crossover. For instance, the BB $111***$ is preferred relative to $11****$. Similarly, the optimal two-schemata $1****1$ is preferred relative to $11****$. Thus, in this case, contrary to the BBH, large BBs or schemata are preferred relative to their lower-order or smaller counterparts.

In Figure 8 we see the analogous graph to Figure 7 but now for a population size of $2^N/10$. We can now see at this population size the dominant effect of drift. However, it is important to notice that generally the curves have a tendency to increase, thus showing that for this landscape recombination is aiding the evolution in finding optimal BBs or schemata.

6   Conclusions

In this paper we have shown how a selecto-recombinative EA can be most naturally understood in terms of Building Blocks that form a basis for the description of the dynamics and which, in distinction to the notion of building block inherent in the BBH and HST, can be given a complete and rigorous mathematical characterization. By examining the exact course grained evolution equations (the equations in the BBB), we saw that one could predict when recombination would be useful and when not and that this depended on both the fitness landscape and the current population state. Further, we noted that, rather than ask whether or not recombination was useful, it was instead more appropriate to ask which recombination masks were useful as a given crossover mask uniquely defined a pair of conjugate BBs.

We showed that recombination was especially benficial in modular landscapes, where there exists a natural notion of landscape "block", i.e. a subset of loci with mutual epistasis, and zero epistasis with loci outside of the subset. The utility of recombination very much depended on having masks that defined BBs that were compatible with these landscape blocks. We considered a family of landscapes with tunable modularity by changing the number of landscape blocks, NIAH and counting ones being the two extremes of this modularity spectrum. For NIAH, save that the optimum is not present, we could see that recombination was not useful. In this sense, the less modular is the landscape the more appropriate it is to use mutation as the principle search operator. The more modular the landscape the more useful recombination potentially becomes, at least if the landscape and recombination blocks are compatible. Hence, for counting ones it is better to have as much recombination as possible, as there do not exist crossover masks that can disrupt the landscape blocks. On the other hand for landscapes with $M < N$ blocks, and hence $M - 1$ crossover points that do not disrupt a landscape block, it is better to have masks that cut only at landscape block boundaries.

There are other questions that equally can be addressed in an analogous fashion, such as: under what conditions mutation might be expected to be more useful than recombination. We can also understand that the recombination distribution is an absolutely fundamental quantity, that specifying it a priori - such as one-point, two-point, uniform crossover etc. - introduces a bias as to what the crossover defined BBs are, and that this bias might be completely inappropriate for the fitness landscape under consideration. In fact, this is precisely the reason why the Royal Road "program" "failed" - there were so many destructive masks (those that cut at points within a landscape block) compared to constructive ones (ones that cut at a landscape block boundary) that recombination was doomed to be suboptimal. The question then is how does one choose suitable masks? Of course, a qualititative understanding of the interaction between landscape and recombination blocks will certainly help. In more complicated cases however, it is probably useful to put the recombination distribution under evolutionary control, i.e., to let it evolve so that the most efficacious (i.e., effectively fit) masks are selected for. It seems to us to be highly likely that that is precisely what has happened in nature and is the explanation as to why recombination "hot spots" have emerged.

Acknowledgments

CRS and JC wish to thank the Macroproyecto - "Tecnologías para la Universidad de la Información y la Computación" for financial support. JC thanks Conacyt for a doctoral fellowship.

References

1. Holland, J.H.: Adaptation in Natural and Artificial Systems. MIT Press, Cambridge, MA (1993)
2. Goldberg, D.E.: Genetic Algorithms in Search, Optimization and Machine Learning. Addison Wesley, Reading, MA (1989)
3. Vose, M.D.: The simple genetic algorithm: Foundations and theory. MIT Press, Cambridge, MA (1999)
4. Mitchell, M., Forrest, S., Holland, J.H.: The royal road for genetic algorithms: Fitness landscapes and ga performance. In: Varela, F.J., Bourgine, P. (eds.) Proceedings of the First European Conference on Artificial Life, pp. 243–254. MIT Press, Cambridge, MA (1992)
5. Forrest, S., Mitchell, M.: Relative building-block fitness and the building-block hypothesis. In: Whitley, L.D. (ed.) Foundations of genetic algorithms 2, pp. 109–126. Morgan Kaufmann, San Mateo (1993)
6. Jansen, T., Wegener, I.: Real royal road functionswhere crossover provably is essential. Discrete Applied Mathematics 149(1-3), 111–125 (2005)
7. Watson, R.A.: Analysis of recombinative algorithms on a non-separable building-block problem. In: Foundations of Genetic Algorithms, pp. 69–89. Morgan Kaufmann, San Mateo (2001)
8. Stephens, C.R., Waelbroeck, H.: Schemata evolution and building blocks. Evol. Comp. 7, 109–124 (1999)

9. Stephens, C.R.: Some exact results from a coarse grained formulation of genetic dynamics. In: Proceedings of the Genetic and Evolutionary Computation Conference (GECCO-2001), July 7-11, 2001, pp. 631–638. Morgan Kaufmann, San Francisco, California, USA (2001)

10. Stephens, C.R., Waelbroeck, H., Aguirre, R.: Schemata as building blocks: Does size matter? In: FOGA, pp. 117–133 (1998)

11. Beyer, H.-G.: An alternative explanation for the manner in which genetic algorithms operate. BioSystems 41(1), 1–15 (1997)

12. Jones, T.: Crossover, macromutation, and population-based search. In: Eshelman, L.J. (ed.) Proceedings of the Sixth International Conference on Genetic Algorithms, pp. 73–80. Morgan Kaufmann, San Francisco, CA (1995)

13. Grefenstette, J.J.: Deception considered harmful. In: Foundations of Genetic Algorithms, pp. 75–91. Morgan Kaufmann, San Mateo (1992)

14. Harik, G.R., Goldberg, D.G.: Learning linkage. In: FOGA 4, pp. 247–262 (1996)

15. Stephens, C.R.: The renormalization group and the dynamics of genetic systems. Acta Phys. Slov. 52, 515–524 (2002)

Sufficient Conditions for Coarse-Graining Evolutionary Dynamics

Keki Burjorjee

DEMO Lab,
Computer Science Department,
Brandeis University, Waltham, MA 02454
`kekib@cs.brandeis.edu`

Abstract. It is commonly assumed that the ability to track the frequencies of a set of schemata in the evolving population of an infinite population genetic algorithm (IPGA) under different fitness functions will advance efforts to obtain a theory of adaptation for the simple GA. Unfortunately, for IPGAs with long genomes and non-trivial fitness functions there do not currently exist theoretical results that allow such a study. We develop a simple framework for analyzing the dynamics of an infinite population evolutionary algorithm (IPEA). This framework derives its simplicity from its abstract nature. In particular we make no commitment to the data-structure of the genomes, the kind of variation performed, or the number of parents involved in a variation operation. We use this framework to derive abstract conditions under which the dynamics of an IPEA can be *coarse-grained*. We then use this result to derive concrete conditions under which it becomes computationally feasible to closely approximate the frequencies of a family of schemata of relatively low order over multiple generations, even when the bitstsrings in the evolving population of the IPGA are long.

1 Introduction

It is commonly assumed that theoretical results which allow one to track the frequencies of schemata in an evolving population of an infinite population genetic algorithm (IPGA) under different fitness functions will lead to a better understanding of how GAs perform adaptation [7,6,8]. An IPGA with genomes of length ℓ can be modelled by a set of 2^ℓ coupled difference equations. For each genome in the search space there is a corresponding state variable which gives the frequency of the genome in the population, and a corresponding difference equation which describes how the value of that state variable in some generation can be calculated from the values of the state variables in the previous generation. A naive way to calculate the frequency of some schema over multiple generations is to numerically iterate the IPGA over many generations, and for each generation, to sum the frequencies of all the genomes that belong to the schema. The simulation of one generation of an IPGA with a genome set of size

C.R. Stephens et al. (Eds.): FOGA 2007, LNCS 4436, pp. 35–53, 2007.
© Springer-Verlag Berlin Heidelberg 2007

N has time complexity $O(N^3)$, and an IPGA with bitstring genomes of length ℓ has a genome set of size $N = 2^\ell$. Hence, the time complexity for a numeric simulation of one generation of an IPGA is $O(8^\ell)$. (See [19, p.36] for a description of how the Fast Walsh Transform can be used to bring this bound down to $O(3^\ell)$.) Even when the Fast Walsh Transform is used, computation time still increases exponentially with ℓ. Therefore for large ℓ the naive way of calculating the frequencies of schemata over multiple generations clearly becomes computationally intractable[1].

Holland's schema theorem [7,6,8] was the first theoretical result which allowed one to calculate (albeit imprecisely) the frequencies of schemata after a single generation. The crossover and mutation operators of a GA can be thought to destroy some schemata and construct others. Holland only considered the destructive effects of these operators. His theorem was therefore an inequality. Later work [15] contained a theoretical result which gives exact values for the schema frequencies after a single generation. Unfortunately for IPGAs with long bitstrings this result does not straightforwardly suggest conditions under which schema frequencies can be numerically calculated over multiple generations in a computationally tractable way.

1.1   The Promise of Coarse-Graining

Coarse-graining is a technique that has widely been used to study aggregate properties (e.g. temperature) of many-body systems with very large numbers of state variables (e.g. gases). This technique allows one to reduce some system of difference or differential equations with many state variables (called the fine-grained system) to a new system of difference or differential equations that describes the time-evolution of a smaller set of state variables (the coarse-grained system). The state variables of the fine-grained system are called the microscopic variables and those of the coarse-grained system are called the macroscopic variables. The reduction is done using a surjective non-injective function between the microscopic state space and the macroscopic state space called the partition function. States in the microscopic state space that share some key property (e.g. energy) are projected to a single state in the macroscopic state space. The reduction is therefore 'lossy', i.e. information about the original system is typically lost. Metaphorically speaking, just as a stationary light bulb projects the *shadow* of some moving 3D object onto a flat 2D wall, the partition function projects the changing state of the fine-grained system onto states in the state space of the coarse-grained system.

The term 'coarse-graining' has been used in the Evolutionary Computation literature to describe different sorts of reductions of the equations of an IPGA. Therefore we now clarify the sense in which we use this term. In this paper a reduction of a system of equations must satisfy three conditions to be called a coarse-graining. Firstly, the number of macroscopic variables should be smaller

---

[1] Vose reported in 1999 that computational concerns force numeric simulation to be limited to cases where $\ell \leq 20$.

than the number of microscopic variables. Secondly, the new system of equations must be completely self-contained in the sense that the state-variables in the new system of equations must *not* be dependent on the microscopic variables. Thirdly, the dynamics of the new system of equations must 'shadow' the dynamics described by the original system of equations in the sense that if the projected state of the original system at time $t = 0$ is equal to the state of the new system at time $t = 0$ then at any other time t, the projected state of the original system should be closely approximated by the state of the new system. If the approximation is instead an equality then the reduction is said to be an *exact* coarse-graining. Most coarse-grainings are not exact. This specification of coarse-graining is consistent with the way this term is typically used in the scientific literature. It is also similar to the definition of coarse-graining given in [12] (the one difference being that in our specification a coarse-graining is assumed not to be exact unless otherwise stated).

Suppose the vector of state variables $\mathbf{x}^{(t)}$ is the state of some system at time t and the vector of state variables $\mathbf{y}^{(t)}$ is the state of a coarse-grained system at time t. Now, if the partition function projects $\mathbf{x}^{(0)}$ to $\mathbf{y}^{(0)}$, then, since none of the state variables of the original system are needed to express the dynamics of the coarse-grained system, one can determine how the state of the coarse-grained system $\mathbf{y}^{(t)}$ (the shadow state) changes over time without needing to determine how the state in the fine-grained system $\mathbf{x}^{(t)}$ (the shadowed state) changes. Thus, even though for any t, one might not be able to determine $\mathbf{x}^{(t)}$, one can always be confident that $\mathbf{y}^{(t)}$ is its projection. Therefore, if the number of state variables of the coarse-grained system is small enough, one can numerically iterate the dynamics of the (shadow) state vector $\mathbf{y}^{(t)}$ without needing to determine the dynamics of the (shadowed) state vector $\mathbf{x}^{(t)}$.

In this paper we give sufficient conditions under which it is possible to coarse-grain the dynamics of an IPGA such that the macroscopic variables are the frequencies of the family of schemata in some schema partition. If the size of this family is small then, regardless of the length of the genome, one can use the coarse-graining result to numerically calculate the approximate frequencies of these schemata over multiple generations in a computationally tractable way. Given some population of bitstring genomes, the set of frequencies of a family of schemata describe the multivariate marginal distribution of the population over the defined locii of the schemata. Thus another way to state our contribution is that we give sufficient conditions under which the multivariate marginal distribution of an evolving population over a small number of locii can be numerically approximated over multiple generations regardless of the length of the genomes.

We stress that our use of the term 'coarse-graining' differs from the way this term has been used in other publications. For instance in [16] the term 'coarse-graining' is used to describe a reduction of the IPGA equations such that each equation in the new system is similar in form to the equations in the original system. The state variables in the new system are defined in terms of the state variables in the original system. Therefore a numerical iteration of the the new system is only computationally tractable when the length of the genomes is relatively short. Elsewhere the term coarse-graining has been defined as "a collection of subsets of the search space that covers the search space"[5], and as "just a function from a genotype set to some other set"[4].

1.2   Some Previous Coarse-Graining Results

Techniques from statistical mechanics have been used to coarse-grain GA dynamics in [9,10,11] (see [13] for a survey of applications of statistical mechanics approaches to GAs). The macroscopic variables of these coarse-grainings are the first few cumulants of the fitness distribution of the evolving population. In [12] several exact coarse-graining results are derived for an IPGA whose variation operation is limited to mutation.

Wright et. al. show in [20] that the dynamics of a non-selective IPGA can be coarse-grained such that the macroscopic variables are the frequencies of a family of schemata in a schema partition. However they argue that the dynamics of a regular selecto-mutato-recombinative IPGA cannot be similarly coarse-grained "except in the trivial case where fitness is a constant for each schema in a schema family"[20]. Let us call this condition *schematic fitness invariance*. Wright et. al. imply that it is so severe that it renders the coarse-graining result essentially useless.

This negative result holds true when there is no constraint on the initial population. In this paper we show that if we constrain the class of initial populations then it is possible to coarse-grain the dynamics of a regular IPGA under a much *weaker* constraint on the fitness function. The constraint on the class of initial populations is not onerous; this class includes the uniform distribution over the genome set.

1.3   Structure of This Paper

The rest of this paper is organized as follows: in the next section we define the basic mathematical objects and notation which we use to model the dynamics of an infinite population *evolutionary* algorithm (IPEA). This framework is very general; we make no commitment to the data-structure of the genomes, the nature of mutation, the nature of recombination , or the number of parents involved in a recombination. We do however require that selection be fitness proportional. In section 3 we define the concepts of semi-coarsenablity, coarsenablity and global coarsenablity which allow us to formalize a useful class of exact coarse-grainings. In section 4 and section 5 we prove some stepping-stone results about selection and variation. We use these results in section 6 where we prove that an IPEA that satisfies certain abstract conditions can be coarse-grained. The proofs in sections 5 and 6 rely on lemmas which have been relegated to and proved in the appendix. In section 7 we specify concrete conditions under which IPGAs with long genomes and non-trivial fitness functions can be coarse-grained such that the macroscopic variables are schema frequencies and the fidelity of the coarse-graining is likely to be high. We conclude in section 8 with a summary of our work.

2   Mathematical Preliminaries

Let X, Y be sets and let $\xi : X \to Y$ be some function. For any $y \in Y$ we use the notation $\langle y \rangle_\xi$ to denote the pre-image of y, i.e. the set $\{x \in X \mid \beta(x) = y\}$. For any subset $A \subset X$ we use the notation $\xi(A)$ to denote the set $\{y \in Y \mid \xi(a) = y$ and $a \in A\}$.

As in [17], for any set X we use the notation Λ^X to denote the set of all distributions over X, i.e. Λ^X denotes set $\{f : X \to [0,1] \mid \sum_{x \in X} f(x) = 1\}$. For any set X, let $0^X : X \to \{0\}$ be the constant zero function over X. For any set X, an m-parent transmission function [14,1,18] over X is an element of the set

$$\left\{ T : \prod_1^{m+1} X \to [0,1] \;\middle|\; \forall x_1, \ldots, x_m \in X, \sum_{x \in X} T(x, x_1', \ldots, x_m') = 1 \right\}$$

Extending the notation introduced above, we denote this set by Λ_m^X. Following [17], we use conditional probability notation in our denotation of transmission functions. Thus an m-parent transmission function $T(x, x_1, \ldots, x_m)$ is denoted $T(x \mid x_1, \ldots, x_m)$.

A transmission function can be used to model the individual-level effect of mutation, which operates on one parent and produces one child, and indeed the individual-level effect of any variation operation which operates on any numbers of parents and produces one child.

Our scheme for modeling EA dynamics is based on the one used in [17]. We model the genomic populations of an EA as distributions over the genome set. The population-level effect of the evolutionary operations of an EA is modeled by mathematical operators whose inputs and outputs are such distributions.

The expectation operator, defined below, is used in the definition of the selection operator, which follows thereafter.

Definition 1. (EXPECTATION OPERATOR) *Let X be some finite set, and let $f : X \to \mathbb{R}^+$ be some function. We define the expectation operator $\mathcal{E}_f : \Lambda^X \cup 0^X \to \mathbb{R}^+ \cup \{0\}$ as follows:*

$$\mathcal{E}_f(p) = \sum_{x \in X} f(x) p(x)$$

The selection operator is parameterized by a fitness function. It models the effect of fitness proportional selection on a population of genomes.

Definition 2. (SELECTION OPERATOR) *Let X be some finite set and let $f : X \to \mathbb{R}^+$ be some function. We define the Selection Operator $\mathcal{S}_f : \Lambda^X \to \Lambda^X$ as follows:*

$$(\mathcal{S}_f p)(x) = \frac{f(x) p(x)}{\mathcal{E}_f(p)}$$

The population-level effect of variation is modeled by the variation operator. This operator is parameterized by a transmission function which models the effect of variation at the individual level.

Definition 3. (VARIATION OPERATOR[2]) *Let X be a countable set, and for any $m \in \mathbb{N}^+$, let $T \in \Lambda_m^X$ be a transmission function over X. We define the variation operator $\mathcal{V}_T : \Lambda^X \to \Lambda^X$ as follows:*

$$(\mathcal{V}_T p)(x) = \sum_{\substack{(x_1,\ldots,x_m) \\ \in \prod_1^m X}} T(x|x_1,\ldots,x_m) \prod_{i=1}^m p(x_i)$$

The next definition describes the projection operator (previously used in [19] and [17]). A projection operator that is parameterized by some function β 'projects' distributions over the domain of β, to distributions over its co-domain.

Definition 4. (PROJECTION OPERATOR) *Let X be a countable set, let Y be some set, and let $\beta : X \to Y$ be a function. We define the projection operator, $\Xi_\beta : \Lambda^X \to \Lambda^Y$ as follows:*

$$(\Xi_\beta p\,)(y) = \sum_{x \in \langle y \rangle_\beta} p(x)$$

and call $\Xi_\beta p$ the β-projection of p.

3   Formalization of a Class of Coarse-Grainings

The following definition introduces some convenient function-related terminology.

Definition 5. (PARTITIONING, THEME SET, THEMES, THEME CLASS) *Let X, K be sets and let $\beta : X \to K$ be a surjective function. We call β a partitioning, call the co-domain K of β the theme set of β, call any element in K a theme of β, and call the pre-image $\langle k \rangle_\beta$ of some $k \in K$, the theme class of k under β.*

The next definition formalizes a class of coarse-grainings in which the macroscopic and microscopic state variables always sum to 1.

Definition 6 (Semi-Coarsenablity, Coarsenablity, Global Coarsenablity). *Let G, K be sets, let $\mathcal{W} : \Lambda^G \to \Lambda^G$ be an operator, let $\beta : G \to K$ be a partitioning, and let $U \subseteq \Lambda^G$ such that $\Xi_\beta(U) = \Lambda^K$. We say that $\mathcal{W}$ is semi-coarsenable under β on U if there exists an operator $\mathcal{Q} : \Lambda^K \to \Lambda^K$ such that for all $p \in U$, $\mathcal{Q} \circ \Xi_\beta p = \Xi_\beta \circ \mathcal{W} p$, i.e. the following diagram commutes:*

$$
\begin{array}{ccc}
U & \xrightarrow{\;\;\mathcal{W}\;\;} & \Lambda^G \\
{\scriptstyle \Xi_\beta}\big\downarrow & & \big\downarrow{\scriptstyle \Xi_\beta} \\
\Lambda^K & \xrightarrow[\;\;\mathcal{Q}\;\;]{} & \Lambda^K
\end{array}
$$

---

[2] Also called the Mixing Operator in [19] and [17].

Since β is surjective, if $\mathcal{Q}$ exists, it is clearly unique; we call it the quotient. We call $G, K, W,$ and U the domain, co-domain, primary operator and turf respectively. If in addition $\mathcal{W}(U) \subseteq U$ we say that $\mathcal{W}$ is coarsenable under β on U. If in addition $U = \Lambda^G$ we say that $\mathcal{W}$ is globally coarsenable under β.

Note that the partition function Ξ_β of the coarse-graining is not the same as the partitioning β of the coarsening.

Global coarsenablity is a stricter condition than coarsenablity, which in turn is a stricter condition than semi-coarsenablity. It is easily shown that global coarsenablity is equivalent to Vose's notion of compatibility [19, p. 188] (for a proof see Theorem 17.5 in [19]).

If some operator $\mathcal{W}$ is coarsenable under some function β on some turf U with some quotient $\mathcal{Q}$, then for any distribution $p_K \in \Xi_\beta(U)$, and all distributions $p_G \in \langle p_K \rangle_{\Xi_\beta}$, one can study the *projected* effect of the repeated application of $\mathcal{W}$ to p_G simply by studying the effect of the repeated application of $\mathcal{Q}$ to p_K. If the size of K is small then a computational study of the projected effect of the repeated application of $\mathcal{W}$ to distributions in U becomes feasible.

4   Global Coarsenablity of Variation

We show that some variation operator $\mathcal{V}_T$ is globally coarsenable under some partitioning if a relationship, that we call *ambivalence*, exists between the transmission function T of the variation operator and the partitioning.

To illustrate the idea of ambivalence consider a partitioning β which partitions a genome set G into three subsets. Fig 1 depicts the behavior of a two-parent transmission function that is ambivalent under β. Given two parents and some child, the probability that the child will belong to some theme class depends *only* on the theme classes of the parents and *not* on the specific parent genomes. Hence the name 'ambivalent' — it captures the sense that when viewed from the coarse-grained level of the theme classes, a transmission function 'does not care' about the specific genomes of the parents or the child.

The definition of ambivalence that follows is equivalent to but more useful than the definition given in [4].

Definition 7. (AMBIVALENCE) *Let G, K be countable sets, let $T \in \Lambda_m^G$ be a transmission function, and let $\beta : G \to K$ be a partitioning. We say that T is ambivalent under β if there exists some transmission function $D \in \Lambda_m^K$, such that for all $k, k_1, \ldots, k_m \in K$ and for any $x_1 \in \langle k_1 \rangle_\beta, \ldots, x_m \in \langle k_m \rangle_\beta$,*

$$\sum_{x \in \langle k \rangle_\beta} T(x|x_1, \ldots, x_m) = D(k|k_1, \ldots, k_m)$$

If such a D exits, it is clearly unique. We denote it by $T^{\vec{\beta}}$ and call it the theme transmission function.

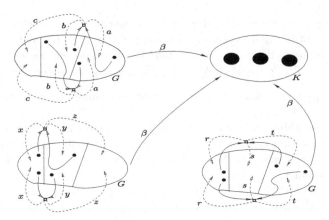

Fig. 1. Let $\beta : G \rightarrow K$ be a coarse-graining which partitions the genome set G into three theme classes. This figure depicts the behavior of a two-parent variation operator that is ambivalent under β. The small dots denote specific genomes and the solid unlabeled arrows denote the recombination of these genomes. A dashed arrow denotes that a child from a recombination may be produced 'somewhere' within the theme class that it points to, and the label of a dashed arrow denotes the probability with which this might occur. As the diagram shows the probability that the child of a variation operation will belong to a particular theme class depends *only* on the theme classes of the parents and *not* on their specific genomes.

Suppose $T \in \Lambda_m^X$ is ambivalent under some $\beta : X \rightarrow K$, we can use the projection operator to express the projection of T under β as follows: for all $k, k_1, \ldots, k_m \in K$, and any $x_1 \in \langle k_1 \rangle_\beta, \ldots, x_m \in \langle k_m \rangle_\beta$, $T^{\vec{\beta}}(k | k_1, \ldots k_m)$ is given by $(\Xi_\beta(T(\cdot | x_1, \ldots, x_m)))(k)$. The notion of ambivalence is equivalent to a generalization of Toussaint's notion of trivial neutrality [17, p. 26]. A one-parent transmission function is ambivalent under a mapping to the set of phenotypes if and only if it is trivially neutral.

The following theorem shows that a variation operator is globally coarsenable under some partitioning if it is parameterized by a transmission function which is ambivalent under that partitioning. The method by which we prove this theorem extends the method used in the proof of Theorem 1.2.2 in [17].

Theorem 1 (Global Coarsenablity of Variation). *Let G and K be countable sets, let $T \in \Lambda_m^G$ be a transmission function and let $\beta : G \rightarrow K$ be some partitioning such that T is ambivalent under β. Then $\mathcal{V}_T : \Lambda^G \rightarrow \Lambda^G$ is globally coarsenable under β with quotient $\mathcal{V}_{T\vec{\beta}}$, i.e. the following diagram commutes:*

$$
\begin{array}{ccc}
\Lambda^G & \xrightarrow{\mathcal{V}_T} & \Lambda^G \\
\Xi_\beta \downarrow & & \downarrow \Xi_\beta \\
\Lambda^K & \xrightarrow[\mathcal{V}_{T\vec{\beta}}]{} & \Lambda^K
\end{array}
$$

PROOF: For any $p \in \Lambda^G$,

$$(\Xi_\beta \circ \mathcal{V}_T p)(k)$$

$$= \sum_{\substack{x \in \langle k \rangle_\beta}} \sum_{\substack{(x_1,\ldots,x_m) \\ \in \prod_1^m X}} T(x|x_1,\ldots,x_m) \prod_{i=1}^m p(x_i)$$

$$= \sum_{\substack{(x_1,\ldots,x_m) \\ \in \prod_1^m X}} \sum_{\substack{x \in \langle k \rangle_\beta}} T(x|x_1,\ldots,x_m) \prod_{i=1}^m p(x_i)$$

$$= \sum_{\substack{(x_1,\ldots,x_m) \\ \in \prod_1^m X}} \prod_{i=1}^m p(x_i) \sum_{x \in \langle k \rangle_\beta} T(x|x_1,\ldots,x_m)$$

$$= \sum_{\substack{(k_1,\ldots,k_m) \\ \in \prod_1^m K}} \sum_{\substack{(x_1,\ldots,x_m) \\ \in \prod_{j=1}^m \langle k_j \rangle_\beta}} \prod_{i=1}^m p(x_i) \sum_{x \in \langle k \rangle_\beta} T(x|x_1,\ldots,x_m)$$

$$= \sum_{\substack{(k_1,\ldots,k_m) \\ \in \prod_1^m K}} \sum_{\substack{(x_1,\ldots,x_m) \\ \in \prod_{j=1}^m \langle k_j \rangle_\beta}} \prod_{i=1}^m p(x_i) T^{\vec{\beta}}(k|k_1,\ldots,k_m)$$

$$= \sum_{\substack{(k_1,\ldots,k_m) \\ \in \prod_1^m K}} T^{\vec{\beta}}(k|k_1,\ldots,k_m) \sum_{\substack{(x_1,\ldots,x_m) \\ \in \prod_{j=1}^m \langle k_j \rangle_\beta}} \prod_{i=1}^m p(x_i)$$

$$= \sum_{\substack{(k_1,\ldots,k_m) \\ \in \prod_1^m K}} T^{\vec{\beta}}(k|k_1,\ldots,k_m) \sum_{x_1 \in \langle k_1 \rangle_\beta} \cdots \sum_{x_m \in \langle k_m \rangle_\beta} p(x_1)\ldots p(x_m)$$

$$= \sum_{\substack{(k_1,\ldots,k_m) \\ \in \prod_1^m K}} T^{\vec{\beta}}(k|k_1,\ldots,k_m) \left(\sum_{x_1 \in \langle k_1 \rangle} p(x_1) \right) \ldots \left(\sum_{x_m \in \langle k_m \rangle} p(x_m) \right)$$

$$= \sum_{\substack{(k_1,\ldots,k_m) \\ \in \prod_1^m K}} T^{\vec{\beta}}(k|k_1,\ldots,k_m) \prod_{i=1}^m \left((\Xi_\beta p)(k_i) \right)$$

$$= (\mathcal{V}_{T^{\vec{\beta}}} \circ \Xi_\beta p)(k) \qquad \qquad \square$$

The implicit parallelism theorem in [20] is similar to the theorem above. Note however that the former theorem only shows that variation is globally coarsenable if firstly, the genome set consists of "fixed length strings, where the size of the alphabet can vary from position to position", secondly the partition over the genome set is a schema partition, and thirdly variation is 'structural' (see [20] for details). The global coarsenablity of variation theorem has none of these specific requirements. Instead it is premised on the existence of an abstract relationship – ambivalence – between the variation operation and a partitioning. The abstract nature of this relationship makes this theorem applicable to evolutionary algorithms other than GAs. In addition this theorem illuminates the essential relationship between 'structural' variation and schemata which was used (implicitly) in the proof of the implicit parallelism theorem.

In [4] it is shown that a variation operator that models any combination of variation operations that are commonly used in GAs — i.e. any combination of mask based crossover and 'canonical' mutation, in any order — is ambivalent under any partitioning that maps bitstrings to schemata (such a partitioning is called a schema partitioning). Therefore 'common' variation in IPGAs is globally coarsenable under *any* schema partitioning. This is precisely the result of the implicit parallelism theorem.

5   Limitwise Semi-coarsenablity of Selection

For some fitness function $f : G \to \mathbb{R}^+$ and some partitioning $\beta : G \to K$ let us say that f is *thematically invariant* under β if, for any schema $k \in K$, the genomes that belong to $\langle k \rangle_\beta$ all have the same fitness. Paraphrasing the discussion in [20] using the terminology developed in this paper, Wright et. al. argue that if the selection operator is globally coarsenable under some schema partitioning $\beta : G \to K$ then the fitness function that parameterizes the selection operator is 'schematically' invariant under β. It is relatively simple to use contradiction to prove a generalization of this statement for arbitrary partitionings.

Schematic invariance is a very strict condition for a fitness function. An IPGA whose fitness function meets this condition is unlikely to yield any substantive information about the dynamics of real world GAs.

As stated above, the selection operator is not *globally* coarsenable unless the fitness function satisfies thematic invariance, however if the set of distributions that selection operates over (i.e. the turf) is appropriately constrained, then, as we show in this section, the selection operator is *semi*-coarsenable over the turf even when the fitness function only satisfies a much *weaker* condition called thematic *mean* invariance.

For any partitioning $\beta : G \to K$, any theme k, and any distribution $p \in \Lambda^G$, the theme conditional operator, defined below, returns a conditional distribution in Λ^G that is obtained by normalizing the probability mass of the elements in $\langle k \rangle_\beta$ by $(\Xi_\beta p)(k)$

Definition 8 (Theme Conditional Operator). *Let G be some countable set, let K be some set, and let $\beta : G \to K$ be some function. We define the theme* conditional operator $\mathcal{C}_\beta : \Lambda^G \times K \to \Lambda^G \cup 0^G$ as follow: For any $p \in \Lambda^G$, and any $k \in K$, $\mathcal{C}_\beta(p,k) \in \Lambda^G \cup 0^G$ such that for any $x \in \langle k \rangle_\beta$,

$$(\mathcal{C}_\beta(p,k))(x) = \begin{cases} 0 & \text{if } (\Xi_\beta p)(k) = 0 \\ \frac{p(x)}{(\Xi_\beta p)(k)} & \text{otherwise} \end{cases}$$

A useful property of the theme conditional operator is that it can be composed with the expected fitness operator to give an operator that returns the average fitness of the genomes in some theme class. To be precise, given some finite genome set G, some partitioning $\beta : G \to K$, some fitness function $f : G \to \mathbb{R}^+$, some distribution $p \in \Lambda^G$, and some theme $k \in K$, $\mathcal{E}_f \circ \mathcal{C}_\beta(p,k)$ is the average fitness of the genomes in $\langle k \rangle_\beta$. This property proves useful in the following definition.

Definition 9 (Bounded Thematic Mean Divergence, Thematic Mean Invariance). *Let G be some finite set, let K be some set, let $\beta : G \to K$ be a partitioning, let $f : G \to \mathbb{R}^+$ and $f^* : K \to \mathbb{R}^+$ be functions, let $U \subseteq \Lambda^G$, and let $\delta \in \mathbb{R}_0^+$. We say that the thematic mean divergence of f with respect to f^* on U under β is bounded by δ if, for any $p \in U$ and for any $k \in K$*

$$|\mathcal{E}_f \circ \mathcal{C}_\beta(p,k) - f^*(k)| \le \delta$$

If $\delta = 0$ we say that f is thematically mean invariant with respect to f^* on U

The next definition gives us a means to measure a 'distance' between real valued functions over finite sets.

Definition 10 (Manhattan Distance Between Real Valued Functions). *Let X be a finite set then for any functions f, h of type $X \to \mathbb{R}$ we define the manhattan distance between f and h, denoted by $d(f,h)$, as follows:*

$$d(f,h) = \sum_{x \in X} |f(x) - h(x)|$$

It is easily checked that d is a metric.

Let $f : G \to \mathbb{R}^+$, $\beta : G \to K$ and $f^* : K \to \mathbb{R}^+$ be functions with finite domains, and let $U \in \Lambda^G$. The following theorem shows that if the thematic mean divergence of f with respect to f^* on U under β is bounded by some δ, then in the limit as $\delta \to 0$, $\mathcal{S}_f$ is semi-coarsenable under β on U .

Theorem 2 (Limitwise Semi-Coarsenablity of Selection). *Let G and K be finite sets, let $\beta : G \to K$ be a partitioning, Let $U \subseteq \Lambda^G$ such that $\Xi_\beta(U) = \Lambda^K$, let $f : G \to \mathbb{R}^+$, $f^* : K \to \mathbb{R}^+$ be some functions such that the thematic mean divergence of f with respect to f^* on U under β is bounded by δ, then for any $p \in U$ and any $\epsilon > 0$ there exists a $\delta' > 0$ such that,*

$$\delta < \delta' \Rightarrow d(\Xi_\beta \circ \mathcal{S}_f p, \mathcal{S}_{f^*} \circ \Xi_\beta p) < \epsilon$$

We depict the result of this theorem as follows:

$$
\begin{array}{ccc}
U & \xrightarrow{\;\mathcal{S}_f\;} & \Lambda^G \\
{\scriptstyle \Xi_\beta}\downarrow & \underset{\delta\to 0}{\lim} & \downarrow{\scriptstyle \Xi_\beta} \\
\Lambda^K & \xrightarrow{\;\mathcal{S}_{f^*}\;} & \Lambda^K
\end{array}
$$

PROOF: For any $p \in U$ and for any $k \in K$,

$$(\Xi_\beta \circ \mathcal{S}_f p)(k)$$

$$= \sum_{g\in\langle k\rangle_\beta} (\mathcal{S}_f p)(g)$$

$$= \sum_{g\in\langle k\rangle_\beta} \frac{f(g).p(g)}{\sum_{g'\in G} f(g').p(g')}$$

$$= \frac{\displaystyle\sum_{g\in\langle k\rangle_\beta} f(g).(\Xi_\beta p)(k).(\mathcal{C}_\beta(p,k))(g)}{\displaystyle\sum_{k'\in K}\sum_{g'\in\langle k'\rangle_\beta} f(g').(\Xi_\beta p)(k')(\mathcal{C}_\beta(p,k'))(g')}$$

$$= \frac{\displaystyle(\Xi_\beta p)(k)\sum_{g\in\langle k\rangle_\beta} f(g).(\mathcal{C}_\beta(p,k))(g)}{\displaystyle\sum_{k'\in K}(\Xi_\beta p)(k')\sum_{g'\in\langle k'\rangle_\beta} f(g').(\mathcal{C}_\beta(p,k'))(g')}$$

$$= \frac{(\Xi_\beta p)(k).\mathcal{E}_f \circ \mathcal{C}_\beta(p,k)}{\displaystyle\sum_{k'\in K}(\Xi_\beta p_G)(k').\mathcal{E}_f \circ \mathcal{C}_\beta(p,k')}$$

$$= (\mathcal{S}_{\mathcal{E}_f\circ\mathcal{C}_\beta(p,\cdot)} \circ \Xi_\beta p)(k)$$

So we have that

$$d(\Xi_\beta \circ \mathcal{S}_f p, \mathcal{S}_{f^*} \circ \Xi_\beta p) = d(\mathcal{S}_{\mathcal{E}_f\circ\mathcal{C}_\beta(p,\cdot)} \circ \Xi_\beta p, \mathcal{S}_{f^*} \circ \Xi_\beta p)$$

By Lemma, 4 (in the appendix) for any $\epsilon > 0$ there exists a $\delta_1 > 0$ such that,

$$d(\mathcal{E}_f \circ \mathcal{C}_\beta(p,.), f^*) < \delta_1 \Rightarrow d(\mathcal{S}_{\mathcal{E}_f\circ\mathcal{C}_\beta(p,\cdot)}(\Xi_\beta p), \mathcal{S}_{f^*}(\Xi_\beta p)) < \epsilon$$

Now, if $\delta < \frac{\delta'}{|K|}$, then $d(\mathcal{E}_f\circ\mathcal{C}_\beta(p,.), f^*) < \delta_1$ □

Corollary 1. *If $\delta = 0$, i.e. if f is thematically mean invariant with respect to f^* on U, then $\mathcal{S}_f$ is semi-coarsenable under β on U with quotient $\mathcal{S}_{f^*}$, i.e. the following diagram commutes:*

$$
\begin{array}{ccc}
U & \xrightarrow{\;\mathcal{S}_f\;} & \Lambda^G \\
{\scriptstyle \Xi_\beta}\downarrow & & \downarrow{\scriptstyle \Xi_\beta} \\
\Lambda^K & \xrightarrow{\;\mathcal{S}_{f^*}\;} & \Lambda^K
\end{array}
$$

6   Limitwise Coarsenablity of Evolution

The two definitions below formalize the idea of an infinite population model of an EA, and its dynamics[3].

Definition 11 (Evolution Machine). *An evolution machine (EM) is a tuple* (G, T, f) *where* G *is some set called the domain,* $f : G \to \mathbb{R}^+$ *is a function called the fitness function and* $T \in \Lambda_m^G$ *is called the transmission function.*

Definition 12 (Evolution Epoch Operator). *Let* $E = (G, T, f)$ *be an evolution machine. We define the evolution epoch operator* $\mathcal{G}_E : \Lambda^G \to \Lambda^G$ *as follows:*

$$\mathcal{G}_E = \mathcal{V}_T \circ \mathcal{S}_f$$

For some evolution machine E, our aim is to give sufficient conditions under which, for any $t \in \mathbb{Z}^+$, $\mathcal{G}_E^t$ approaches coarsenablity in the limit. The following definition gives us a formal way to state one of these conditions.

Definition 13 (Non-Departure). *Let* $E = (G, T, f)$ *be an evolution machine, and let* $U \subseteq \Lambda^G$. *We say that* E *is* non-departing *over* U *if*

$$\mathcal{V}_T \circ \mathcal{S}_f(U) \subseteq U$$

Note that our definition does *not* require $\mathcal{S}_f(U) \subseteq U$ in order for E to be non-departing over U.

Theorem 3 (Limitwise Coarsenablity of Evolution). *Let* $E = (G, T, f)$, *be an evolution machine such that* G *is finite, let* $\beta : G \to K$ *be some partitioning, let* $f^* : K \to \mathbb{R}^+$ *be some function, let* $\delta \in \mathbb{R}_0^+$, *and let* $U \subseteq \Lambda^G$ *such that* $\Xi_\beta(U) = \Lambda^K$. *Suppose that the following statements are true:*

1. *The thematic mean divergence of* f *with respect to* f^* *on* U *under* β *is bounded by* δ
2. T *is ambivalent under* β
3. E *is non-departing over* U

Then, letting $E^* = (K, T^{\vec{\beta}}, f^*)$ *be an evolution machine, for any* $t \in \mathbb{Z}^+$ *and any* $p \in U$,

1. $\mathcal{G}_E^t p \in U$
2. *For any* $\epsilon > 0$, *there exists* $\delta' > 0$ *such that,*

$$\delta < \delta' \Rightarrow d(\Xi_\beta \circ \mathcal{G}_E^t p \, , \, \mathcal{G}_{E^*}^t \circ \Xi_\beta p) < \epsilon$$

---

[3] The definition of an EM given here is different from its definition in [2,3]. The fitness function in this definition maps genomes directly to fitness values. It therefore subsumes the genotype-to-phenotype and the phenotype-to-fitness functions of the previous definition. In previous work these two functions were always composed together; their subsumption within a single function increases clarity.

We depict the result of this theorem as follows:

$$
\begin{array}{ccc}
U & \xrightarrow{\;\mathcal{G}_E^t\;} & U \\
{\scriptstyle\Xi_\beta}\big\downarrow & {\scriptstyle\lim_{\delta\to 0}} & \big\downarrow{\scriptstyle\Xi_\beta} \\
\Lambda^K & \xrightarrow[\;\mathcal{G}_{E*}^t\;]{} & \Lambda^K
\end{array}
$$

PROOF: We prove the theorem for any $t \in \mathbb{Z}_0^+$. The proof is by induction on t. The base case, when $t = 0$, is trivial. For some $n = \mathbb{Z}_0^+$, let us assume the hypothesis for $t = n$. We now show that it is true for $t = n + 1$. For any $p \in U$, by the inductive assumption $\mathcal{G}_E^n p$ is in U. Therefore, since E is non-departing over U, $\mathcal{G}_E^{n+1} p \in U$. This completes the proof of the first part of the hypothesis. For a proof of the second part note that,

$$
d(\Xi_\beta \circ \mathcal{G}_E^{n+1} p , \, \mathcal{G}_{E*}^{n+1} \circ \Xi_\beta p)
$$
$$
= d(\Xi_\beta \circ \mathcal{V}_T \circ \mathcal{S}_f \circ \mathcal{G}_E^n p, \, \mathcal{V}_{T^{\vec\beta}} \circ \mathcal{S}_{f*} \circ \mathcal{G}_{E*}^n \circ \Xi_\beta p)
$$
$$
= d(\mathcal{V}_{T^{\vec\beta}} \circ \Xi_\beta \circ \mathcal{S}_f \circ \mathcal{G}_E^n p, \, \mathcal{V}_{T^{\vec\beta}} \circ \mathcal{S}_{f*} \circ \mathcal{G}_{E*}^n \circ \Xi_\beta p) \qquad \text{(by theorem 1)}
$$

Hence, for any $\epsilon > 0$, by Lemma 2 there exists δ_1 such that

$$
d(\Xi_\beta \circ \mathcal{S}_f \circ \mathcal{G}_E^n p, \, \mathcal{S}_{f*} \circ \mathcal{G}_{E*}^n \circ \Xi_\beta p) < \delta_1 \Rightarrow d(\Xi_\beta \circ \mathcal{G}_E^{n+1} p, \, \mathcal{G}_{E*}^{n+1} \circ \Xi_\beta p) < \epsilon
$$

As d is a metric it satisfies the triangle inequality. Therefore we have that

$$
d(\Xi_\beta \circ \mathcal{S}_f \circ \mathcal{G}_E^n p, \, \mathcal{S}_{f*} \circ \mathcal{G}_{E*}^n \circ \Xi_\beta p) \leq
$$
$$
d(\Xi_\beta \circ \mathcal{S}_f \circ \mathcal{G}_E^n p, \, \mathcal{S}_{f*} \circ \Xi_\beta \circ \mathcal{G}_E^n p) +
$$
$$
d(\mathcal{S}_{f*} \circ \Xi_\beta \circ \mathcal{G}_E^n p, \, \mathcal{S}_{f*} \circ \mathcal{G}_{E*}^n \circ \Xi_\beta p)
$$

By our inductive assumption $\mathcal{G}_E^n p \in U$. So, by theorem 2 there exists a δ_2 such that

$$
\delta < \delta_2 \Rightarrow d(\Xi_\beta \circ \mathcal{S}_f \circ \mathcal{G}_E^n p, \, \mathcal{S}_{f*} \circ \Xi_\beta \circ \mathcal{G}_E^n p) < \frac{\delta_1}{2}
$$

By lemma 3 there exists a δ_3 such that

$$
d(\Xi_\beta \circ \mathcal{G}_E^n p, \, \mathcal{G}_{E*}^n \circ \Xi_\beta p) < \delta_3 \Rightarrow d(\mathcal{S}_{f*} \circ \Xi_\beta \circ \mathcal{G}_E^n p, \, \mathcal{S}_{f*} \circ \mathcal{G}_{E*}^n \circ \Xi_\beta p) < \frac{\delta_1}{2}
$$

By our inductive assumption, there exists a δ_4 such that

$$
\delta < \delta_4 \Rightarrow d(\Xi_\beta \circ \mathcal{G}_E^n p, \, \mathcal{G}_{E*}^n \circ \Xi_\beta p) < \delta_3
$$

Therefore, letting $\delta' = \min(\delta_2, \delta_4)$ we get that

$$
\delta < \delta^* \Rightarrow d(\Xi_\beta \circ \mathcal{G}_E^{n+1} p, \mathcal{G}_{E*}^{n+1} \circ \Xi_\beta p) < \epsilon \qquad\qquad \square
$$

The limitwise coarsenability of evolution theorem is very general. As we have not committed ourselves to any particular genomic data-structure the coarse-graining result we have obtained is applicable to any IPEA provided that it satisfies three abstract conditions: bounded thematic mean divergence, ambivalence, and non-departure. The fidelity of the coarse-graining depends on the the the minimal bound on the thematic mean divergence. Maximum fidelity is achieved in the limit as this minimal bound tends to zero.

7 Sufficient Conditions for Coarse-Graining IPGA Dynamics

We now use the result in the previous section to argue that the dynamics of an IPGA with long genomes, uniform crossover, and fitness proportional selection can be coarse-grained with high fidelity for a relatively coarse schema partitioning, provided that the initial population satisfies a constraint called *approximate achematic uniformity* and the fitness function satisfies a constraint called *low-variance schematic fitness distribution*. We stress at the outset that our argument is principled but informal, i.e. though the argument rests relatively straightforwardly on theorem 3, we do find it necessary in places to appeal to the reader's intuitive understanding of GA dynamics.

For any $n \in \mathbb{Z}^+$, let $\mathfrak{B}_n$ be the set of all bitstrings of length n. For some $\ell \gg 1$ and some $m \ll \ell$, let $\beta : \mathfrak{B}_\ell \to \mathfrak{B}_m$ be some schema partitioning. Let $f^* : \mathfrak{B}_m \to \mathbb{R}^+$ be some function. For each $k \in \mathfrak{B}_m$, let $D_k \in \Lambda^{\mathbb{R}^+}$ be some distribution over the reals with low variance such that the mean of distribution D_k is $f^*(k)$. Let $f : \mathfrak{B}_\ell \to \mathbb{R}^+$ be a fitness function such that for any $k \in \mathfrak{B}_m$, the fitness values of the elements of $\langle k \rangle_\beta$ are independently drawn from the distribution D_k. For such a fitness function we say that fitness is *schematically distributed with low-variance*.

Let U be a set of distributions such that for any $k \in \mathfrak{B}_m$ and any $p \in U$, $\mathcal{C}_\beta(p, k)$ is approximately uniform. It is easily checked that U satisfies the condition $\Xi_\beta(U) = \Lambda^{\mathfrak{B}_m}$. We say that the distributions in U are *approximately schematically uniform*.

Let δ be the minimal bound such that for all $p \in U$ and for all $k \in \mathfrak{B}_m$, $|\mathcal{E}_f \circ \mathcal{C}_\beta(p, k) - f^*(k)| \le \delta$. Then, for any $\epsilon > 0$, $\mathbf{P}(\delta < \epsilon) \to 1$ as $\ell - m \to \infty$. Because we have chosen ℓ and m such that $\ell - m$ is 'large', it is reasonable to assume that the minimal bound on the schematic mean divergence of f on U under β is likely to be 'low'.

Let $T \in \Lambda^{\mathfrak{B}_\ell}$ be a transmission function that models the application of uniform crossover. In sections 6 and 7 of [4] we rigorously prove that a transmission function that models any mask based crossover operation is ambivalent under any schema partitioning. Uniform crossover is mask based, and β is a schema partitioning, therefore T is ambivalent under β.

Let $p_{\frac{1}{2}} \in \Lambda^{\mathfrak{B}_1}$ be such that $p_{\frac{1}{2}}(0) = \frac{1}{2}$ and $p_{\frac{1}{2}}(1) = \frac{1}{2}$. For any $p \in U$, $\mathcal{S}_f p$ may be 'outside' U because there may be one or more $k \in \mathfrak{B}_m$ such that $\mathcal{C}_\beta(\mathcal{S}_f p, k)$ is not quite uniform. Recall that for any $k \in \mathfrak{B}_m$ the variance of D_k is low. Therefore even though $\mathcal{S}_f p$ may be 'outside' U, the deviation from schematic uniformity is not likely to be large. Furthermore, given the low variance of D_k, the marginal distributions of $\mathcal{C}_\beta(\mathcal{S}_f p, k)$ will be very close to $p_{\frac{1}{2}}$. Given these facts and our choice of transmission function, for all $k \in K$, $\mathcal{C}_\beta(\mathcal{V}_T \circ \mathcal{S}_f p, k)$ will be more uniform than $\mathcal{C}_\beta(\mathcal{S}_f p, k)$, and we can assume that $\mathcal{V}_T \circ \mathcal{S}_f p$ is in U. In other words, we can assume that E is non-departing over U.

Let $E = (\mathfrak{B}_\ell, T, f)$ and $E^* = (\mathfrak{B}_m, T^{\vec{\beta}}, f^*)$ be evolution machines. By the discussion above and the limitwise coarsenablity of evolution theorem one can expect that for any approximately thematically uniform distribution $p \in U$ (including of course the uniform distribution over $\mathfrak{B}_\ell$), the dynamics of E^* when initialized with $\Xi_\beta p$ will approximate the projected dynamics of E when initialized with p. As the bound δ is 'low', the fidelity of the approximation will be 'high'.

Note that the constraint that fitness be low-variance schematically distributed, which is required for this coarse-graining, is much weaker than the very strong constraint of schematic fitness invariance (all genomes in each schema must have the *same* value) which is required to coarse-grain IPGA dynamics in [20].

8   Conclusion

It is commonly assumed that the ability to track the frequencies of schemata in an evolving infinite population across multiple generations under different fitness functions will lead to better theories of adaptation for the simple GA. Unfortunately tracking the frequencies of schemata in the naive way described in the introduction is computationally intractable for IPGAs with long genomes. A previous coarse-graining result [20] suggests that tracking the frequencies of a family of low order schemata is computationally feasible, regardless of the length of the genomes, if fitness is schematically invariant (with respect to the family of schemata). Unfortunately this strong constraint on the fitness function renders this result useless if one's goal is to understand how GAs perform adaptation on real-world fitness functions.

In this paper we developed a simple yet powerful abstract framework for modeling evolutionary dynamics. We used this framework to show that the dynamics of an IPEA can be coarse-grained if it satisfies three abstract conditions. We then used this result to argue that the evolutionary dynamics of an IPGA with fitness proportional selection and uniform crossover can be coarse-grained (with high fidelity) under a relatively coarse schema partitioning if the initial distribution satisfies a constraint called approximate schematic uniformity (a very reasonable condition), and fitness is low-variance schematically distributed. The latter condition is much weaker than the schematic invariance constraint previously required to coarse-grain selecto-mutato-recombinative evolutionary dynamics.

Acknowledgements. The reviewers of this paper gave me many useful comments, suggestions, and references. I thank them for their feedback. I also thank Jordan Pollack for supporting this work.

References

1. Altenberg, L.: The evolution of evolvability in genetic programming. In: Kinnear, Jr., K.E. (ed.) Advances in Genetic Programming, MIT Press, Cambridge, MA (1994)
2. Burjorjee, K., Pollack, J.B.: Theme preservation and the evolution of representation. In: Theory of Representation Workshop, GECCO (2005)
3. Burjorjee, K., Pollack, J.B.: Theme preservation and the evolution of representation. In: IICAI, pp. 1444–1463 (2005)
4. Burjorjee, K., Pollack, J.B.: A general coarse-graining framework for studying simultaneous inter-population constraints induced by evolutionary operations. In: GECCO 2006. Proceedings of the 8th annual conference on Genetic and evolutionary computation, ACM Press, New York (2006)
5. Contreras, A.A., Rowe, J.E., Stephens, C.R.: Coarse-graining in genetic algorithms: Some issues and examples. In: Cantú-Paz, E., Foster, J.A., Deb, K., Davis, L., Roy, R., O'Reilly, U.-M., Beyer, H.-G., Kendall, G., Wilson, S.W., Harman, M., Wegener, J., Dasgupta, D., Potter, M.A., Schultz, A., Dowsland, K.A., Jonoska, N., Miller, J., Standish, R.K. (eds.) GECCO 2003. LNCS, vol. 2724, pp. 874–885. Springer, Heidelberg (2003)
6. Goldberg, D.E.: Genetic Algorithms in Search, Optimization & Machine Learning. Addison-Wesley, Reading, MA (1989)
7. Holland, J.H.: Adaptation in Natural and Artificial Systems: An Introductory Analysis with Applications to Biology, Control, and Artificial Intelligence. University of Michigan (1975)
8. Mitchell, M.: An Introduction to Genetic Algorithms. The MIT Press, Cambridge, MA (1996)
9. Prügel-Bennet, A., Shapiro, J.L.: An analysis of genetic algorithms using statistical mechanics. Phys. Rev. Lett. 72(9), 1305 (1994)
10. Prügel-Bennett, A., Shapiro, J.L.: The dynamics of a genetic algorithm for the ising spin-glass chain. Physica D 104, 75–114 (1997)
11. Rattray, M., Shapiro, J.L.: Cumulant dynamics of a population under multiplicative selection, mutation, and drift. Theoretical Population Biology 60, 17–32 (2001)
12. Rowe, J.E., Vose, M.D., Wright, A.H.: Differentiable coarse graining. Theor. Comput. Sci 361(1), 111–129 (2006)
13. Shapiro, J.L.: Statistical mechanics theory of genetic algorithms. In: Kallel, L., Naudts, B., Rogers, A. (eds.) Theoretical Aspects of Evolutionary Computing, pp. 87–108. Springer, Heidelberg (2001)
14. Slatkin, M.: Selection and polygenic characters. PNAS 66(1), 87–93 (1970)
15. Stephens, C.R., Waelbroeck, H.: Effective degrees of freedom in genetic algorithms and the block hypothesis. In: ICGA, pp. 34–40 (1997)
16. Stephens, C.R., Zamora, A.: EC theory: A unified viewpoint. In: Cantú-Paz, E., Foster, J.A., Deb, K., Davis, L., Roy, R., O'Reilly, U.-M., Beyer, H.-G., Kendall, G., Wilson, S.W., Harman, M., Wegener, J., Dasgupta, D., Potter, M.A., Schultz, A., Dowsland, K.A., Jonoska, N., Miller, J., Standish, R.K. (eds.) GECCO 2003. LNCS, vol. 2724, Springer, Heidelberg (2003)
17. Toussaint, M.: The Evolution of Genetic Representations and Modular Neural Adaptation. PhD thesis, Institut fr Neuroinformatik, Ruhr-Universiät-Bochum, Germany (2003)
18. Toussaint, M.: On the evolution of phenotypic exploration distributions. In: Foundations of Genetic Algorithms 7 (FOGA VII), Morgan Kaufmann, San Francisco (2003)
19. Vose, M.D.: The simple genetic algorithm: foundations and theory. MIT Press, Cambridge (1999)

20. Wright, A.H., Vose, M.D., Rowe, J.E.: Implicit parallelism. In: Cantú-Paz, E., Foster, J.A., Deb, K., Davis, L., Roy, R., O'Reilly, U.-M., Beyer, H.-G., Kendall, G., Wilson, S.W., Harman, M., Wegener, J., Dasgupta, D., Potter, M.A., Schultz, A., Dowsland, K.A., Jonoska, N., Miller, J., Standish, R.K. (eds.) GECCO 2003. LNCS, vol. 2724, Springer, Heidelberg (2003)

Appendix

Lemma 1. *For any finite set X, and any metric space (Υ, d), let $\mathcal{A} : \Upsilon \to \Lambda^X$ and let $\mathcal{B} : X \to [\Upsilon \to [0,1]]$ be functions[4] such that for any $h \in \Upsilon$, and any $x \in X$, $(\mathcal{B}(x))(h) = (\mathcal{A}(h))(x)$. For any $x \in X$, and for any $h^* \in \Upsilon$, if the following statement is true*

$$\forall x \in X, \forall \epsilon_x > 0, \exists \delta_x > 0, \forall h \in \Upsilon, d(h, h^*) < \delta_x \Rightarrow |(\mathcal{B}(x))(h) - (\mathcal{B}(x))(h^*)| < \epsilon_x$$

Then we have that

$$\forall \epsilon > 0, \exists \delta > 0, \forall h \in \Upsilon, d(h, h^*) < \delta \Rightarrow d(\mathcal{A}(h), \mathcal{A}(h^*)) < \epsilon$$

This lemma says that $\mathcal{A}$ is continuous at h^* if for all $x \in X$, $\mathcal{B}(x)$ is continuous at h^*. PROOF: We first prove the following two claims

Claim 1

$$\forall x \in X \text{ s.t. } (\mathcal{B}(x))(h^*) > 0, \forall \epsilon_x > 0, \exists \delta_x > 0, \forall h \in \Upsilon,$$
$$d(h, h^*) < \delta_x \Rightarrow |(\mathcal{B}(x))(h) - (\mathcal{B}(x))(h^*)| < \epsilon_x.(\mathcal{B}(x))(h^*)$$

This claim follows from the continuity of $\mathcal{B}(x)$ at h^* for all $x \in X$ and the fact that $(\mathcal{B}(x))(h^*)$ is a positive constant w.r.t. h.

Claim 2 *For all $h \in \Upsilon$*

$$\sum_{\substack{x \in X \text{ s.t.} \\ (\mathcal{A}(h^*))(x) > \\ (\mathcal{A}(h))(x)}} |(\mathcal{A}(h^*))(x) - (\mathcal{A}(h))(x)| = \sum_{\substack{x \in X \text{ s.t.} \\ (\mathcal{A}(h))(x) > \\ (\mathcal{A}(h^*))(x)}} |(\mathcal{A}(h))(x) - (\mathcal{A}(h^*))(x)|$$

The proof of this claim is as follows: for all $h \in \Upsilon$,

$$\sum_{x \in X} (\mathcal{A}(h^*)(x)) - (\mathcal{A}(h))(x) = 0$$

$$\Rightarrow \sum_{\substack{x \in X \text{ s.t.} \\ (\mathcal{A}(h^*))(x) > \\ (\mathcal{A}(h))(x)}} (\mathcal{A}(h^*))(x) - (\mathcal{A}(h))(x) - \sum_{\substack{x \in X \text{ s.t.} \\ (\mathcal{A}(h))(x) > \\ (\mathcal{A}(h^*))(x)}} (\mathcal{A}(h))(x) - (\mathcal{A}(h^*))(x) = 0$$

$$\Rightarrow \sum_{\substack{x \in X \text{ s.t.} \\ (\mathcal{A}(h^*))(x) > \\ (\mathcal{A}(h))(x)}} (\mathcal{A}(h^*))(x) - (\mathcal{A}(h))(x) = \sum_{\substack{x \in X \text{ s.t.} \\ (\mathcal{A}(h))(x) > \\ (\mathcal{A}(h^*))(x)}} (\mathcal{A}(h))(x) - (\mathcal{A}(h^*))(x)$$

$$\Rightarrow \left| \sum_{\substack{x \in X \text{ s.t.} \\ (\mathcal{A}(h^*))(x) > \\ (\mathcal{A}(h))(x)}} (\mathcal{A}(h^*))(x) - (\mathcal{A}(h))(x) \right| = \left| \sum_{\substack{x \in X \text{ s.t.} \\ (\mathcal{A}(h))(x) > \\ (\mathcal{A}(h^*))(x)}} (\mathcal{A}(h))(x) - (\mathcal{A}(h^*))(x) \right|$$

---

[4] For any sets X, Y we use the notation $[X \to Y]$ to denote the set of all functions from X to Y.

$$\Rightarrow \sum_{\substack{x \in X \text{s.t.} \\ (\mathcal{A}(h^*))(x) > \\ (\mathcal{A}(h))(x)}} |(\mathcal{A}(h^*))(x) - (\mathcal{A}(h))(x)| = \sum_{\substack{x \in X \text{s.t.} \\ (\mathcal{A}(h))(x) > \\ (\mathcal{A}(h^*))(x)}} |(\mathcal{A}(h))(x) - (\mathcal{A}(h^*))(x)|$$

We now prove the lemma. Using claim 1 and the fact that X is finite, we get that $\forall \epsilon > 0, \exists \delta > 0, \forall h \in [X \to \mathbb{R}]$ such that $d(h, h^*) < \delta$,

$$\sum_{\substack{x \in X \text{s.t.} \\ (\mathcal{A}(h^*))(x) > \\ (\mathcal{A}(h))(x)}} |(\mathcal{B}(x))(h^*) - (\mathcal{B}(x))(h)| < \sum_{\substack{x \in X \text{s.t.} \\ (\mathcal{A}(h^*))(x) > \\ (\mathcal{A}(h))(x)}} \frac{\epsilon}{2} \cdot (\mathcal{B}(x))(h^*)$$

$$\Rightarrow \sum_{\substack{x \in X \text{s.t.} \\ (\mathcal{A}(h^*))(x) > \\ (\mathcal{A}(h))(x)}} |(\mathcal{A}(h^*))(x) - (\mathcal{A}(h))(x)| < \sum_{\substack{x \in X \text{s.t.} \\ (\mathcal{A}(h^*))(x) > \\ (\mathcal{A}(h))(x)}} \frac{\epsilon}{2} \cdot (\mathcal{A}(h^*))(x)$$

$$\Rightarrow \sum_{\substack{x \in X \text{s.t.} \\ (\mathcal{A}(h^*))(x) > \\ (\mathcal{A}(h))(x)}} |(\mathcal{A}(h^*))(x) - (\mathcal{A}(h))(x)| < \frac{\epsilon}{2} \qquad \square$$

By Claim 2 and the result above, we have that $\forall \epsilon > 0, \exists \delta > 0, \forall h \in [X \to \mathbb{R}]$ such that $d(h, h^*) < \delta$,

$$\sum_{\substack{x \in X \text{s.t.} \\ (\mathcal{A}(h))(x) > \\ (\mathcal{A}(h^*))(x)}} |(\mathcal{A}(h))(x) - (\mathcal{A}(h^*))(x)| < \frac{\epsilon}{2}$$

Therefore, given the two previous results, we have that $\forall \epsilon > 0, \exists \delta > 0, \forall h \in [X \to \mathbb{R}]$ such that $d(h, h^*) < \delta$,

$$\sum_{x \in X} |(\mathcal{A}(h))(x) - (\mathcal{A}(h^*)(x))| < \epsilon \qquad \square$$

Lemma 2. *Let X be a finite set, and let $T \in \Lambda_m^X$ be a transmission function. Then for any $p' \in \Lambda^X$ and any $\epsilon > 0$, there exists a $\delta > 0$ such that for any $p \in \Lambda^X$,*

$$d(p, p') < \delta \Rightarrow d(\mathcal{V}_T p, \mathcal{V}_T p') < \epsilon$$

Sketch of Proof: Let $\mathcal{A} : \Lambda^X \to \Lambda^X$ be defined such that $(\mathcal{A}(p))(x) = (\mathcal{V}_T p)(x)$. Let $\mathcal{B} : X \to [\Lambda^X \to [0, 1]]$ be defined such that $(\mathcal{B}(x))(p) = (\mathcal{V}_T p)(x)$. The reader can check that for any $x \in X$, $\mathcal{B}(x)$ is a continuous function. The application of lemma 1 completes the proof.

By similar arguments, we obtain the following two lemmas.

Lemma 3. *Let X be a finite set, and let $f : X \to \mathbb{R}^+$ be a function. Then for any $p' \in \Lambda^X$ and any $\epsilon > 0$, there exists a $\delta > 0$ such that for any $p \in \Lambda^X$,*

$$d(p, p') < \delta \Rightarrow d(\mathcal{S}_f p, \mathcal{S}_f p') < \epsilon$$

Lemma 4. *Let X be a finite set, and let $p \in \Lambda^X$ be a distribution. Then for any $f' \in [X \to \mathbb{R}^+]$, and any $\epsilon > 0$, there exists a $\delta > 0$ such that for any $f \in [X \to \mathbb{R}^+]$,*

$$d(f, f') < \delta \Rightarrow d(\mathcal{S}_f p, \mathcal{S}_{f'} p) < \epsilon$$

On the Brittleness of Evolutionary Algorithms

Thomas Jansen

FB 4, LS 2, Universität Dortmund
44221 Dortmund, Germany
Thomas.Jansen@udo.edu

Abstract. Evolutionary algorithms are randomized search heuristics that are often described as robust general purpose problem solvers. It is known, however, that the performance of an evolutionary algorithm may be very sensitive to the setting of some of its parameters. A different perspective is to investigate changes in the expected optimization time due to small changes in the fitness landscape. A class of fitness functions where the expected optimization time of the (1+1) evolutionary algorithm is of the same magnitude for almost all of its members is the set of linear fitness functions. Using linear functions as a starting point, a model of a fitness landscape is devised that incorporates important properties of linear functions. Unexpectedly, the expected optimization time of the (1+1) evolutionary algorithm is clearly larger for this fitness model than on linear functions.

1 Introduction

Evolutionary algorithms (EAs) belong to the broad class of general randomized search heuristics. They are popular because they are easy to implement, easy to apply to different kinds of problems, and because they are believed to be robust. It is known, however, that the performance of evolutionary algorithms can be very sensitive to relatively small changes in the algorithm or the settings of their parameters. For example, Storch [15] proved that for some simple mutation-based evolutionary algorithms even changes of the population size by only 1 can lead to enormous changes in the performance. This brittleness does not hinge on the evolutionary algorithm to be complex in any way. Even for the very simple (1+1) evolutionary algorithm seemingly small changes can lead to extreme performance changes. Usually, in the (1+1) evolutionary algorithm, an offspring replaces its parent if its fitness is at least as good. Changing this to "is strictly better" can increase the expected optimization time from a small polynomial to exponential [12]. Clearly, the choice of the mutation probability can have an enormous influence, too [11].

Here, we take a different perspective. We consider the influence of seemingly small changes of the fitness landscape on an EA's performance. We concentrate on changes that do not directly aim at changing the algorithm's performance. For example, this rules our moving the global optimum drastically [5]. We use the expected optimization time as actual measure of performance. The analysis of evolutionary algorithms with respect to the expected optimization time

C.R. Stephens et al. (Eds.): FOGA 2007, LNCS 4436, pp. 54–69, 2007.

corresponds to the analysis of (randomized) algorithms with respect to the (expected) run time. Whereas the latter is a central aspect in the field of design and analysis of algorithms, the former is still a relative knew branch of evolutionary algorithm theory. It has been acknowledged as an approach of growing importance and has the potential of linking the EA community to the field of design and analysis of (randomized) algorithms. Considering the past fifteen years of analysis of the expected optimization time (beginning with the analysis of the expected optimization time of the (1+1) evolutionary algorithm on ONEMAX [14]) it becomes apparent that impressive progress has been made. In particular, there are various analytical methods developed that greatly simplify the task of analyzing the optimization time of evolutionary algorithms. Here, we present asymptotically tight upper and lower bounds on the expected optimization time with relatively simple proofs.

It is well known that the expected optimization time of the (1+1) EA is $O(n \log n)$ for any linear function and $\Theta(n \log n)$ for linear function that depend on $\Theta(n^\varepsilon)$ bits (for any constant $\varepsilon > 0$) [4]. The lower bound $\Omega(n \log n)$ follows easily from a simple proof of this lower bound for a wide range of evolutionary algorithms and fitness functions [10]. The first rigorous proof of the upper bound was quite complicated. It is based on an abstract but flexible method that is in general difficult to apply. It consists of abstracting from the actual evolutionary algorithm and fitness function and considering a more abstract random process that approximates crucial random variables of the true underlying random process. Clearly, the errors introduced by the approximation have to be proven to be bounded. This approach is similar in spirit to the approach using potential functions in the classical analysis of algorithms and data structures [1]. Now, a much simpler proof using drift analysis is known [9]. Drift analysis allows the derivation of the expected optimization time by the analysis of the expected change of some measure of progress in one generation. It is particularly strong in situations where the expected change in a single generation remains more or less unchanged during the complete run (like, e. g., for LEADINGONES [10]). For such functions, drift analysis can be used to prove asymptotically tight upper and lower bounds on the expected optimization time. But even for functions where this expected change varies quite dramatically (like, e. g., for ONEMAX [10]) it can lead to asymptotically tight upper bounds.

As we already mentioned, our object of interest is the (1+1) EA, a very simple evolutionary algorithm. For the sake of completeness we give a precise definition together with a short introduction of the notions and notations used in the next section. The main part of the paper contains a discussion of common properties of linear functions. We design a fitness landscape that can be described as a pessimistic model of these common properties. We present a formal definition and connect this model to the analysis of the (1+1) EA on linear functions (Sect. 3). The analysis of this model leads to asymptotically tight lower and upper bounds on the expected optimization time. We accompany these theoretically derived asymptotic bounds by empirical data from actual runs. This yields some insight in the quality of the analytic results (Sect. 4). Finally, we summarize our findings in the conclusions.

2   Definitions and Notation

The only evolutionary algorithm considered here is the so-called (1+1) EA. It uses a population size of only 1, an offspring population size of only 1, and, as a consequence, mutation only. The selection for replacement is the deterministic plus-selection known from evolution strategies. It is sometimes considered as a kind of randomized hill-climber [13]. Due to the standard bit mutation employed, it is more accurately described as one of the simplest evolutionary algorithms.

Definition 1 ((1+1) evolutionary algorithm ((1+1) EA))

1. *Initialization*
   $t := 0$. *Choose* $x_t \in \{0,1\}^n$ *uniformly at random.*
2. *Mutation*
   $y := x_t$. *Independently for each bit in* y, *flip this bit with probability* $1/n$.
3. *Selection*
   If $f(y) \geq f(x_t)$ *Then* $x_{t+1} := y$ *Else* $x_{t+1} := x_t$
4. $t := t + 1$. *Continue at line 2.*

The (1+1) EA as described above can be used for maximization of any pseudo-Boolean function $f \colon \{0,1\}^n \to \mathbb{R}$. In the context of optimization, the most interesting question is concerned with the amount of time needed. We call the number of rounds the (1+1) EA needs to optimize f the optimization time T, i.e.,

$$T = \min \left\{ t \mid f(x_t) = \max \left\{ f(x) \mid x \in \{0,1\}^n \right\} \right\}.$$

For a bit string $x \in \{0,1\}^n$ we refer to the i-th bit in x by $x[i]$. Thus, we have $x = x[1]x[2] \cdots x[n]$. We use the notation b^i for the concatenation of i b-bits ($i \in \mathbb{N}_0$, $b \in \{0,1\}$). For example, we have $1110000 = 1^3 0^4$. Often, we are interested in the number of 1-bits in a bit string x. Clearly, this coincides with the function value of x under the well-known fitness function ONEMAX since $\text{ONEMAX}(x) = \sum_{i=1}^{n} x[i]$.

Each pseudo-Boolean fitness function $f \colon \{0,1\}^n \to \mathbb{R}$ has a unique representation as a polynomial $f(x) = \sum_{I \subseteq \{1,2,\ldots,n\}} w_I \cdot \prod_{i \in I} x[i]$. We call $\deg(f) := \max\{|I| \mid w_I \neq 0\}$ the degree of f. A fitness function $f \colon \{0,1\}^n \to \mathbb{R}$ is called linear, if $\deg(f) = 1$ holds. Clearly, linear functions can be written as $f(x) = w_0 + \sum_{i=1}^{n} w_i \cdot x[i]$ with weights $w_i \in \mathbb{R}$.

For our modeling, we define a partial order on bit strings extending the order on bits in a natural way.

$$\forall x, y \in \{0,1\}^n \colon x \leq y :\Leftrightarrow \forall i \in \{1,\ldots,n\} \colon x[i] \leq y[i]$$

Clearly, this partial order has 0^n as unique minimal element and 1^n as unique maximal element. We write $x \not\leq y$ for the negation of $x \leq y$. Since we are dealing with a partial order this is different from $y \leq x$.

3   Design and Analysis of a Pessimistic Fitness Model for Linear Functions

We consider the (1+1) EA on linear functions $f(x) = w_0 + \sum_{i=1}^{n} w_i$. Due to the selection employed, the additive constant w_0 has no influence on the run of the algorithm. Therefore, we assume $w_0 = 0$ in the following. Since we want to derive an upper bound on the expected optimization time, we may assume without loss of generality that $w_i \neq 0$ holds for all weights. For positions i with $w_i = 0$ the value of $x[i]$ has no influence on the function value. Clearly, this can only decrease the expected optimization time in comparison with non-zero weights.

Each linear function with only non-zero weights has a unique global optimum $x^* \in \{0, 1\}^n$. Since we maximize f, we have $x^*[i] = 1$ for $w_i > 0$ and $x^*[i] = 0$ for $w_i < 0$. The (1+1) EA is completely symmetric with respect to the roles of 0-bits and 1-bits. Therefore, we may exchange 0-bits and 1-bits at arbitrary positions (and change w_i to $-w_i$ accordingly) without changing the behavior of the algorithm. Thus, we can w. l. o. g. restrict our attention to linear functions with only positive weights, i. e., $w_i > 0$ for all $i > 0$.

Linear functions with only positive weights have 1^n as their unique global optimum. For such functions one may measure the progress the (1+1) EA makes on f by considering the random process $\text{ONEMAX}(x_0)$, $\text{ONEMAX}(x_1)$, ... on $\{0, 1, \ldots, n\}$. This is a random sequence of numbers that (depending on the specific linear function f used as a fitness function) may be increasing and decreasing. But for all functions with 1^n as global optimum (not only linear ones), the optimization time coincides with the minimal index t such that $\text{ONEMAX}(x_t) = n$ holds. If one was to use drift analysis and consider the number of 1-bits as measure, one would estimate the optimization time by finding bounds on the expected change in the number of 1-bits in one step, i. e., giving bounds on

$$\text{E}\left(\text{ONEMAX}(x_i) - \text{ONEMAX}(x_{i-1}) \mid x_{i-1}\right).$$

It is worth mentioning that the number of 1-bits in the current bit string may be totally misleading as a measure of progress even for fitness functions that have 1^n as their unique global optimum. Consider for example $\text{PLATEAU}: \{0, 1\}^n \to \mathbb{R}$. The formal definition

$$\text{PLATEAU}(x) := \begin{cases} 2n & \text{if } x = 1^n \\ n + 1 & \text{if } x \in \left\{1^i 0^{n-i} \mid i \in \{1, 2, \ldots, n - 1\}\right\} \\ n - \text{ONEMAX}(x) & \text{otherwise} \end{cases}$$

reveals that the number of 1-bits is decreasing with increasing fitness values for almost all points in the search space. Moreover, it is not difficult to see that even when approaching the optimum on the plateau, i. e. $x_t = 1^i 0^{n-i}$, the expected change in the number of 1-bits is slightly negative. Nevertheless, the (1+1) EA optimizes this function efficiently [12].

For linear functions with only positive weights, however, we expect the number of 1-bits to be a quite accurate measure of progress. Consider randomized local search (RLS), a search heuristic that works just like the (1+1) EA but flips exactly one bit in each mutation step. If a 1-bit flips (becoming a 0-bit), the fitness decreases and the offspring is rejected. If a 0-bit flips (becoming a 1-bit), the fitness increases and the offspring replaces its parent. Thus, the expected increase in the number of 1-bits in one step equals z/n for a bit string with z 0-bits and is strictly positive for all non-optimal points in the search space. Moreover, we get $\mathrm{E}\,(T) \leq \sum_{z=1}^{n} \frac{n}{z} = O(n \log n)$ as an immediate consequence. The (1+1) EA, however, may flip several bits in a single mutation. But mutations of single bits are the most likely mutations to cause changes. Therefore we may expect that the (1+1) EA behaves quite similarly. And, in fact, we know that RLS and the (1+1) EA both have expected optimization time $\Theta(n \log n)$ on the class of linear functions [3,4,9].

Clearly, there are infinitely many different linear functions with only positive weights. Two extreme examples are ONEMAX and BINVAL with $\mathrm{BINVAL}(x) = \sum_{i=1}^{n} 2^{n-i}x[i]$. While for ONEMAX there is a clear correspondence between fitness values and the number of 1-bits, this is not the case for BINVAL. In particular, 10^{n-1} and 01^{n-1} differ in function value by only 1, yet the difference in the number of 1-bits equals $n - 2$. We see that for ONEMAX the number of 1-bits can never decrease during a run whereas it may for BINVAL.

In order to capture the different properties of linear functions we introduce a random process that shares many properties with the (1+1) EA operating on some linear fitness function but that is different in a certain way. We consider this random process (denoted as PO-EA, short for partially ordered EA, since its main ingredient is the partial ordering on $\{0,1\}^n$) in the following as an abstraction of the (1+1) EA operating on linear functions with only positive weights. Note that it is in some sense a pessimistic modeling.

Definition 2 (PO-EA)

1. **Initialization**
   $t := 0$. Choose $x_t \in \{0,1\}^n$ uniformly at random.
2. **Mutation**
   $y := x_t$. Independently for each bit in y, flip this bit with probability $1/n$.
3. **Selection**
   If $(y \geq x) \vee ((y \not\leq x) \wedge (\mathrm{ONEMAX}(y) \leq \mathrm{ONEMAX}(x)))$
   Then $x_{t+1} := y$ Else $x_{t+1} := x_t$
4. $t := t + 1$. Continue at line 2.

We observe that there is no fitness function $f: \{0,1\}^n \to \mathbb{R}$ where the (1+1) EA operates as PO-EA does. Consider, e. g., $n = 4$, $x_0 = 0101$, $x_1 = 0010$, $x_2 = 0110$, $x_3 = 0001$, $x_4 = x_0$. For PO-EA and all $t \in \{0, 1, 2, 3\}$ the transition from x_t

to x_{t+1} is done and, moreover, one of the inequalities is strict. The transition from x_0 to x_1 decreases the number of 1-bits, the transition from x_1 to x_2 flips exactly one 0-bit, the transition from x_2 to x_3 decreases the number of 1-bits again and, finally, the transition from x_3 to x_4 flips exactly one 0-bit. And yet we have $x_0 = x_4$ and the current population cycled. This is only possible for the (1+1) EA with fitness functions if the function values are all equal. But equal function values imply that all the transitions could occur in reversed order. This, however, is clearly not the case here for PO-EA. In fact, none of the four transitions can be directly reversed. In this sense PO-EA operates on an abstract model of a fitness landscape inspired by linear functions but not on a fitness landscape defined by any real fitness function.

Since PO-EA does not operate on a fitness function, we cannot really say that it optimizes. However, there is a unique bit string, 1^n, that cannot be replaced by any offspring. Since this bit string can be reached in one single mutation from any $x \in \{0,1\}^n$, we see that the Markov chain describing PO-EA has 1^n as unique absorbing state. This resembles optimization with the (1+1) EA and therefore we call the expected absorption time the expected optimization time, too. So, the number of steps PO-EA needs to reach 1^n is called its optimization time.

Even though PO-EA acts differently than the (1+1) EA on any fitness function, it resembles the (1+1) EA on a linear fitness function with only positive weights. If only 0-bits are mutated, y replaces its parent. If only 1-bits are mutated, y is discarded. For these two "pure" cases PO-EA and the (1+1) EA agree. For all other "mixed" cases, the offspring replaces its parent in PO-EA only if the number of 1-bits does not increase. This is a worst case behavior with respect to the number of 1-bits as measure of progress. It is worth mentioning that PO-EA is not overly pessimistic in the following sense. Consider the (1+1) EA on BINVAL and assume $x_t = 01^{n-1}$. In this specific situation, PO-EA and the (1+1) EA behave exactly the same. Thus, there is a situation where PO-EA models the (1+1) EA on a specific linear function accurately. Since it is pessimistic in other situations and never optimistic, one may use PO-EA as a model to derive an upper bound on the expected optimization time of the (1+1) EA on linear functions. In fact, this is almost exactly the approach taken by He and Yao in [7]. Due to an error in the calculations, the proof there is not correct. This is "repaired" [8] by considering the (1+1) EA with a mutation probability of only $1/(2n)$. Interestingly, this simplifies the proof dramatically. But this "trick" is dissatisfactory since this is not the most recommended and most often used mutation probability. The result, however, holds for the common mutation probability $1/n$, too, but the proof is more involved [3,4,9]. In the following section, we derive asymptotically tight lower and upper bounds on the expected optimization time of PO-EA.

4   Analysis of PO-EA

The main tool for the analysis of PO-EA's expected optimization time here is drift analysis. Interestingly, different distance measures turn out to be useful for the lower and upper bound. When considering PO-EA as a pessimistic model of the (1+1) EA on linear functions, the most pressing question is whether the modeling is too pessimistic. This question can only be answered affirmatively by means of a lower bound that is $\omega(n \log n)$. Therefore, we start with the presentation of a lower bound. We prove a lower bound of $\Omega(n^{3/2})$. This demonstrates that any pessimistic approach that is equivalent to PO-EA cannot deliver an asymptotically tight bound for the performance of the (1+1) EA on the class of linear functions.

Theorem 1. *The expected optimization time of PO-EA is* $\Omega(n\sqrt{n})$.

Proof. We want to apply drift analysis and use a very simple and straightforward distance measure: the number of 0-bits. We start with an upper bound on the expected increase in the number of 1-bits in one generation. Let $\Delta(t) := \text{ONEMAX}(x_{t+1}) - \text{ONEMAX}(x_t)$. We present a simple derivation of the drift, $E\left(\Delta(t) \mid \text{ONEMAX}(x_t) = n - z\right)$, based on an inspection of all possible mutations of x_t. A mutation is described by the number of mutating 0-bits b_0 and the number of mutating 1-bits b_1. Clearly, the probability of one such mutation equals

$$\binom{z}{b_0} \cdot \binom{n-z}{b_1} \cdot \left(\frac{1}{n}\right)^{b_0+b_1} \cdot \left(1 - \frac{1}{n}\right)^{n-b_0-b_1}$$

and we see that

$$\sum_{b_0=0}^{z} \sum_{b_1=0}^{n-z} \binom{z}{b_0} \cdot \binom{n-z}{b_1} \cdot \left(\frac{1}{n}\right)^{b_0+b_1} \cdot \left(1 - \frac{1}{n}\right)^{n-b_0-b_1} = 1$$

holds since we are dealing with a distribution. The contribution of one such mutation to $E\left(\Delta(t) \mid \text{ONEMAX}(x_t) = n - z\right)$ equals $b_0 - b_1$ if the offspring replaces its parent and 0 otherwise. We know that the offspring replaces its parent if either only 0-bits flip, i.e., $b_1 = 0$, or the number of flipping 1-bits is at least as large as the number of flipping 0-bits, i.e., $b_0 \leq b_1$. Since mutations where the offspring replaces its parent but $b_0 = b_1$ holds have a contribution of 0 to $E\left(\Delta(t) \mid \text{ONEMAX}(x_t) = n - z\right)$, the following holds when we adopt the convention that sums where the lower summation bound exceeds the upper summation bounds contribute 0.

$$E\left(\Delta(t) \mid \text{ONEMAX}(x_t) = n - z\right) = \left(\sum_{b_0=1}^{z} b_0 \binom{z}{b_0} \left(\frac{1}{n}\right)^{b_0} \left(1 - \frac{1}{n}\right)^{n-b_0}\right)$$

$$+ \sum_{b_0=1}^{z} \sum_{b_1=b_0+1}^{n-z} (b_0 - b_1) \binom{z}{b_0} \cdot \binom{n-z}{b_1} \cdot \left(\frac{1}{n}\right)^{b_0+b_1} \cdot \left(1 - \frac{1}{n}\right)^{n-b_0-b_1}$$

We observe that

$$\sum_{b_0=1}^{z} \sum_{b_1=1}^{b_0} (b_0 - b_1) \binom{z}{b_0} \binom{n-z}{b_1} \left(\frac{1}{n}\right)^{b_0+b_1} \left(1 - \frac{1}{n}\right)^{n-b_0-b_1} \geq 0$$

holds. Thus, we obtain an upper bound if we consider the following.

$$\sum_{b_0=1}^{z}\sum_{b_1=0}^{n-z}(b_0-b_1)\binom{z}{b_0}\binom{n-z}{b_1}\left(\frac{1}{n}\right)^{b_0+b_1}\left(1-\frac{1}{n}\right)^{n-b_0-b_1}$$

It is easy to see that

$$\sum_{b_0=1}^{z}\sum_{b_1=0}^{n-z}(b_0-b_1)\binom{z}{b_0}\binom{n-z}{b_1}\left(\frac{1}{n}\right)^{b_0+b_1}\left(1-\frac{1}{n}\right)^{n-b_0-b_1}$$

$$=\sum_{b_0=1}^{z}\sum_{b_1=0}^{n-z}b_0\binom{z}{b_0}\binom{n-z}{b_1}\left(\frac{1}{n}\right)^{b_0+b_1}\left(1-\frac{1}{n}\right)^{n-b_0-b_1}$$

$$-\sum_{b_0=1}^{z}\sum_{b_1=0}^{n-z}b_1\binom{z}{b_0}\binom{n-z}{b_1}\left(\frac{1}{n}\right)^{b_0+b_1}\left(1-\frac{1}{n}\right)^{n-b_0-b_1}$$

$$=\sum_{b_0=0}^{z}\sum_{b_1=0}^{n-z}b_0\binom{z}{b_0}\binom{n-z}{b_1}\left(\frac{1}{n}\right)^{b_0+b_1}\left(1-\frac{1}{n}\right)^{n-b_0-b_1}$$

$$-\sum_{b_0=0}^{z}\sum_{b_1=0}^{n-z}b_1\binom{z}{b_0}\binom{n-z}{b_1}\left(\frac{1}{n}\right)^{b_0+b_1}\left(1-\frac{1}{n}\right)^{n-b_0-b_1}$$

$$+\sum_{b_1=0}^{n-z}b_1\binom{n-z}{b_1}\left(\frac{1}{n}\right)^{b_1}\left(1-\frac{1}{n}\right)^{n-b_1}$$

$$=\frac{z}{n}-\frac{n-z}{n}+\sum_{b_1=0}^{n-z}b_1\binom{n-z}{b_1}\left(\frac{1}{n}\right)^{b_1}\left(1-\frac{1}{n}\right)^{n-b_1}$$

holds, since

$$\sum_{b_0=0}^{z}\sum_{b_1=0}^{n-z}b_0\binom{z}{b_0}\binom{n-z}{b_1}\left(\frac{1}{n}\right)^{b_0+b_1}\left(1-\frac{1}{n}\right)^{n-b_0-b_1}$$

equals the expected number of mutating 0-bits in a bit string with exactly z 0-bits which is z/n and, analogously,

$$\sum_{b_0=0}^{z}\sum_{b_1=0}^{n-z}b_1\binom{z}{b_0}\binom{n-z}{b_1}\left(\frac{1}{n}\right)^{b_0+b_1}\left(1-\frac{1}{n}\right)^{n-b_0-b_1}$$

equals the expected number of mutation 1-bits in a bit string with exactly z 0-bits (and, therefore, $n-z$ 1-bits) which is $(n-z)/n$. The expected values are easy to see since the random variables are obviously binomially distributed with parameters $z, 1/n$ and $n-z, 1/n$ respectively.

We have

$$\sum_{b_1=0}^{n-z}b_1\binom{n-z}{b_1}\left(\frac{1}{n}\right)^{b_1}\left(1-\frac{1}{n}\right)^{n-b_1}$$

$$= \left(1 - \frac{1}{n}\right)^{z} \sum_{b_1=0}^{n-z} b_1 \binom{n-z}{b_1} \left(\frac{1}{n}\right)^{b_1} \left(1 - \frac{1}{n}\right)^{n-z-b_1}$$

and consider a bit string of length $n-z$ that is subject to mutation with mutation probability $1/n$. Clearly, the expected number of mutation bits equals

$$\sum_{b_1=0}^{n-z} b_1 \binom{n-z}{b_1} \left(\frac{1}{n}\right)^{b_1} \left(1 - \frac{1}{n}\right)^{n-z-b_1} = \frac{n-z}{n}$$

and we obtain

$$\frac{z}{n} - \frac{n-z}{n} + \left(1 - \frac{1}{n}\right)^{z} \frac{n-z}{n} = \frac{1}{n} \cdot \left(2z - n + \left(1 - \frac{1}{n}\right)^{z}(n-z)\right)$$

for our upper bound. Application of the binomial theorem yields

$$\frac{1}{n} \cdot \left(2z - n + \left(1 - \frac{1}{n}\right)^{z}(n-z)\right)$$

$$= \frac{1}{n} \left(2z - n + (n-z)\left(1 - \frac{z}{n} + \frac{z(z-1)}{2n^2} + \sum_{i=3}^{z}\binom{z}{i}\left(-\frac{1}{n}\right)^{i}\right)\right) = \Theta\left(\frac{z^2}{n^2}\right)$$

for our upper bound.

For the derivation of a lower bound we want to subtract from our upper bound what we added to the exact expectation. So, we see that

$$E\left(\Delta(t) \mid \text{ONEMAX}(x_t) = n - z\right)$$

$$= \frac{1}{n} \cdot \left(2z - n + \left(1 - \frac{1}{n}\right)^{z}(n-z)\right)$$

$$- \sum_{b_0=1}^{z} \sum_{b_1=1}^{b_0-1} (b_0 - b_1)\binom{z}{b_0}\binom{n-z}{b_1}\left(\frac{1}{n}\right)^{b_0+b_1}\left(1 - \frac{1}{n}\right)^{n-b_0-b_1}$$

holds. We have

$$\sum_{b_0=1}^{z} \sum_{b_1=1}^{b_0-1} (b_0 - b_1)\binom{z}{b_0}\binom{n-z}{b_1}\left(\frac{1}{n}\right)^{b_0+b_1}\left(1 - \frac{1}{n}\right)^{n-b_0-b_1}$$

$$= \sum_{b_0=2}^{z} \sum_{b_1=1}^{b_0-1} (b_0 - b_1)\binom{z}{b_0}\binom{n-z}{b_1}\left(\frac{1}{n}\right)^{b_0+b_1}\left(1 - \frac{1}{n}\right)^{n-b_0-b_1}$$

$$= \frac{z(n-z)}{n^3}\left(1 - \frac{1}{n}\right)^{n-3}$$

$$+ \sum_{b_0=3}^{z} \sum_{b_1=1}^{b_0-1} (b_0 - b_1)\binom{z}{b_0}\binom{n-z}{b_1}\left(\frac{1}{n}\right)^{b_0+b_1}\left(1 - \frac{1}{n}\right)^{n-b_0-b_1}$$

$$= o\left(\frac{z^2}{n^2}\right)$$

and can conclude that

$$E\left(\Delta(t) \mid \text{OneMax}(x_t) = n - z\right) = \Theta\left(\frac{z^2}{n^2}\right) - o\left(\frac{z^2}{n^2}\right)$$

holds. Thus, the bound $\Theta(z^2/n^2)$ is tight.

Now we consider PO-EA. We neglect the first steps after random initialization and start to pay attention when the number of 0-bits is decreased to $\sqrt{n}$ for the first time. As long as the number of 0-bits is $O(\sqrt{n})$, the drift is bounded above by $O(1/n)$. Moreover, the probability to increase the number 0-bits is smaller than the probability to decrease the number of 0-bits. It is easy to see that the probability that in this situation the number of 0-bits increases to $\omega(\sqrt{n})$ is very small. As long as this does not happen, the process has drift $O(1/n)$ and initial distance to the optimum of $\sqrt{n}$. This yields a lower bound of $\Omega(n\sqrt{n})$ for the expected optimization time.   □

We see that PO-EA is significantly slower than the (1+1) EA on any linear function. The finding that the expected change in the number of 0-bits in one generation with currently z 0-bits is $\Theta(z^2/n^2)$ may lead us to believe that the actual expected optimization time is $\Theta(n^2)$. This however, is not correct as the following theorem shows. In fact, the lower bound from Theorem 1 is asymptotically tight.

Theorem 2. *The expected optimization time of PO-EA is $\Theta(n\sqrt{n})$.*

Proof. We use the result on the expected change in the number of 0-bits from the proof of Theorem 1 as a starting point. When the current search point contains z 0-bits, the expected change in the number of 0-bits in one generation is $\Theta(z^2/n^2)$. We can use this result to obtain an upper bound on the expected time needed to decrease the number of 0-bits by at least 1. Clearly, the distance that we need to overcome is 1. If we increase the number of 0-bits in this process, the drift is only increased. Thus, $\Omega(z^2/n^2)$ is a lower bound on the drift and we obtain $O(n^2/z^2)$ as upper bound on the expected time.

This reasoning can easily be extended to an upper bound. We simply add up all these expected waiting times for $z \in \{1, 2, \ldots, n\}$ and obtain

$$\sum_{z=1}^{n} \Theta\left(\frac{n^2}{z^2}\right) = \Theta(n^2)$$

as upper bound. This, however, is clearly larger than the upper bound we want to prove. It is worth mentioning, however, that we can generalize this result in the following way. If we are interested in the first point of time when the number of 0-bits is decreased to at most z', we only need to sum up all expected waiting times for $z \in \{z' + 1, z' + 2, \ldots, n\}$. Then we obtain

$$\sum_{z=z'+1}^{n} \Theta\left(\frac{n^2}{z^2}\right) = \Theta(n^2/z')$$

as upper bound for this waiting time.

Using the number of 0-bits as distance measure, we cannot obtain an asymptotically tight upper bound on PO-EA's expected optimization time. In order to apply drift analysis, we use a different distance measure here, namely $d(x) := \sqrt{n - \text{ONEMAX}(x)}$.

We use drift analysis and consider a measure of distance d that depends on $\text{ONEMAX}(x)$, only. Thus, we may restrict ourselves to some $\tilde{d}: \{0, 1, \ldots, n\} \to \mathbb{R}_0^+$ with $\tilde{d}(n - \text{ONEMAX}(x)) = d(x)$. Therefore, we have $\tilde{d}(z) = 0 \Leftrightarrow z = 0$ and $\tilde{d}$ strictly increasing with z. Since we use $d(x) = \sqrt{n - \text{ONEMAX}(x)}$ as measure for the distance to the global optimum, we have $\tilde{d}(z) = \sqrt{z}$. Since the value of $\tilde{d}$ can easily be computed when $x \in \{0, 1\}^n$ is known, we allow ourselves to use the notation $\tilde{d}(x)$ when we are really referring to $\tilde{d}(n - \text{ONEMAX}(x))$.

We need a lower bound on the expected change of $\tilde{d}$ in one generation. Let $\mathbf{1}_{\text{expr}}$ denote the indicator function such that $\mathbf{1}_{\text{expr}} = \begin{cases} 1 \text{ if expr is true} \\ 0 \text{ otherwise} \end{cases}$ holds. Then we have

$$\text{E}\left(\tilde{d}(x_t) - \tilde{d}(x_{t+1}) \mid \tilde{d}(x_t) = z\right)$$

$$= \sum_{b_0=0}^{z} \sum_{b_1=0}^{n-z} \left(\binom{z}{b_0} \cdot \binom{n-z}{b_1} \cdot \left(\frac{1}{n}\right)^{b_0+b_1} \cdot \left(1 - \frac{1}{n}\right)^{n-b_0-b_1} \right.$$

$$\left. \cdot \mathbf{1}_{(b_1=0<b_0)\vee(b_1>b_0>0)} \cdot \left(\tilde{d}(z) - \tilde{d}(z + b_1 - b_0)\right) \right)$$

since this mirrors all ways of mutating b_0 0-bits and b_1 1-bits and sums up all differences that are accepted by PO-EA. Adopting the notation that $\sum_{i=a}^{b} f(i) = 0$ holds for $a > b$, we obtain

$$\text{E}\left(\tilde{d}(x_t) - \tilde{d}(x_{t+1}) \mid \tilde{d}(x_t) = z\right) = \left(1 - \frac{1}{n}\right)^n \cdot \sum_{b_0=1}^{z} \left(\binom{z}{b_0} \cdot \left(\frac{1}{n-1}\right)^{b_0} \right.$$

$$\left. \cdot \left(\sqrt{z} - \sqrt{z - b_0} + \sum_{b_1=b_0+1}^{n-z} \binom{n-z}{b_1} \cdot \left(\frac{1}{n-1}\right)^{b_1} \cdot \left(\sqrt{z} - \sqrt{z + b_1 - b_0}\right) \right) \right)$$

by dividing the double-sum into two sums, one for the case of only mutating 0-bits, the other for the case of mutating 0-bits and 1-bits.

We consider the differences of square roots $\sqrt{z} - \sqrt{z - b}$ with $0 \le b \le z$. Clearly,

$$\sqrt{z} - \sqrt{z - b} = \frac{b}{\sqrt{z}} \cdot \frac{1}{1 + \sqrt{1 - b/z}}$$

holds and $b/(2\sqrt{z}) \le \sqrt{z} - \sqrt{z - b} \le b/\sqrt{z}$ follows. Moreover, we have $\lim_{z \to \infty} \left(\sqrt{z} - \sqrt{z - b}\right) = b/\left(2\sqrt{z}\right)$ for any b with $b = o(z)$.

We want to estimate

$$\text{E}\left(\tilde{d}(x_t) - \tilde{d}(x_{t+1}) \mid \tilde{d}(x_t) = z\right) = \left(1 - \frac{1}{n}\right)^n \cdot \sum_{b_0=1}^{z} \left(\binom{z}{b_0} \cdot \left(\frac{1}{n-1}\right)^{b_0} \right.$$

$$\cdot \left(\sqrt{z} - \sqrt{z - b_0} + \sum_{b_1 = b_0 + 1}^{n-z} \binom{n-z}{b_1} \cdot \left(\frac{1}{n-1} \right)^{b_1} \cdot \left(\sqrt{z} - \sqrt{z + b_1 - b_0} \right) \right) \right)$$

and are content with an asymptotic result. Thus, we may safely ignore the term $(1 - 1/n)^n$. This leads us to the following calculations.

$$\sum_{b_0=1}^{z} \left(\binom{z}{b_0} \cdot \left(\frac{1}{n-1} \right)^{b_0} \right.$$

$$\left. \cdot \left(\sqrt{z} - \sqrt{z - b_0} - \sum_{b_1 = b_0 + 1}^{n-z} \left(\sqrt{z + b_1 - b_0} - \sqrt{z} \right) \cdot \binom{n-z}{b_1} \cdot \left(\frac{1}{n-1} \right)^{b_1} \right) \right)$$

$$= \sum_{b_0=1}^{z} \left(\binom{z}{b_0} \cdot \left(\frac{1}{n-1} \right)^{b_0} \cdot \left(\frac{b_0}{\sqrt{z}} \cdot \frac{1}{1 + \sqrt{1 - b_0/z}} \right. \right.$$

$$\left. \left. - \sum_{b_1 = b_0 + 1}^{n-z} \frac{b_1 - b_0}{\sqrt{z}} \cdot \frac{1}{1 + \sqrt{1 + (b_1 - b_0)/z}} \cdot \binom{n-z}{b_1} \cdot \left(\frac{1}{n-1} \right)^{b_1} \right) \right)$$

$$= \frac{1}{\sqrt{z}} \cdot \sum_{b_0=1}^{z} \left(\binom{z}{b_0} \cdot \left(\frac{1}{n-1} \right)^{b_0} \right.$$

$$\left. \cdot \left(\frac{b_0}{1 + \sqrt{1 - b_0/z}} - \sum_{b_1 = b_0 + 1}^{n-z} \frac{b_1 - b_0}{1 + \sqrt{1 + (b_1 - b_0)/z}} \cdot \binom{n-z}{b_1} \cdot \left(\frac{1}{n-1} \right)^{b_1} \right) \right)$$

$$= \frac{1}{\sqrt{z}} \cdot \sum_{b_0=1}^{z} \left(\binom{z}{b_0} \cdot \left(\frac{1}{n-1} \right)^{b_0} \right.$$

$$\left. \cdot \left(\frac{b_0}{1 + \sqrt{1 - b_0/z}} - \sum_{j=1}^{n-z-b_0} \frac{j}{1 + \sqrt{1 + j/z}} \cdot \binom{n-z}{b_0 + j} \cdot \left(\frac{1}{n-1} \right)^{b_0 + j} \right) \right)$$

Using $\binom{a}{b} \leq a^b / b!$ to obtain an estimate we get

$$\frac{1}{\sqrt{z}} \cdot \sum_{b_0=1}^{z} \left(\binom{z}{b_0} \cdot \left(\frac{1}{n-1} \right)^{b_0} \right.$$

$$\left. \cdot \left(\frac{b_0}{1 + \sqrt{1 - b_0/z}} - \sum_{j=1}^{n-z-b_0} \frac{j}{1 + \sqrt{1 + j/z}} \cdot \frac{1}{(b_0 + j)!} \cdot \left(\frac{n-z}{n-1} \right)^{b_0 + j} \right) \right)$$

as lower bound. To simplify things let us assume for the moment that

$$\frac{b_0}{1 + \sqrt{1 - b_0/z}} - \sum_{j=1}^{n-z-b_0} \frac{j}{1 + \sqrt{1 + j/z}} \cdot \frac{1}{(b_0 + j)!} \cdot \left(\frac{n-z}{n-1} \right)^{b_0 + j} = \Omega(1) \quad (1)$$

holds. Then we see that a lower bound of order

$$\Omega \left(\frac{1}{\sqrt{z}} \cdot \sum_{b_0=1}^{z} \left(\binom{z}{b_0} \cdot \left(\frac{1}{n-1} \right)^{b_0} \right) \right) = \Omega \left(\frac{\sqrt{z}}{n} \right)$$

follows. Without making use of our assumption (1) we obtain

$$\Omega\left(\frac{\sqrt{z}}{n}\cdot\left(\frac{1}{1+\sqrt{1-1/z}}-\sum_{j=1}^{n-z-1}\frac{j}{1+\sqrt{1+j/z}}\cdot\frac{1}{(j+1)!}\cdot\left(\frac{n-z}{n-1}\right)^{j+1}\right)\right)$$

$$=\Omega\left(\frac{\sqrt{z}}{n}\cdot\left(\frac{1}{1+\sqrt{1-1/z}}-\frac{1}{1+\sqrt{1+1/z}}\cdot\left(1-\frac{z}{n}\right)\cdot\sum_{j=1}^{\infty}\frac{j}{(j+1)!}\right)\right)$$

$$=\Omega\left(\frac{\sqrt{z}}{n}\cdot\left(\frac{1}{1+\sqrt{1-1/z}}-\frac{1}{1+\sqrt{1+1/z}}\cdot\left(1-\frac{z}{n}\right)\right)\right)$$

$$=\Omega\left(\frac{\sqrt{z}}{n}\cdot\left(\frac{1}{z}+\frac{z}{n}\right)\right)=\Omega\left(\frac{1}{n\sqrt{z}}+\frac{z\sqrt{z}}{n^2}\right)$$

as lower bound. This expression becomes asymptotically minimal for $z=\Theta(\sqrt{n})$ leading us to $\Omega\left(n^{-5/4}\right)$ as lower bound. Together with the upper bound $O(\sqrt{n})$ on the maximal distance this leads to an upper bound of $O(n^{7/4})$ for the expected optimization time by drift analysis which is still not tight.

Now we can remember the result we obtained at the beginning of this proof using the number of 0-bits as distance measure. We have that the expected time needed to decrease the number of 0-bits to at most $\sqrt{n}$ for the first time is bounded above by $O(n^2/\sqrt{n})=O(n^{3/2})$. Then, now using $\sqrt{n-\text{ONEMAX}(x)}$ as distance measure again, there is a distance of $n^{1/4}$ to the global optimum. Together with $\Omega(n^{-5/4})$ as lower bound on the drift we obtain $O(n^{6/4})=O(n^{3/2})$ as upper bound on the expected optimization time as claimed. $\Box$

One may wonder why $\sqrt{n-\text{ONEMAX}(x)}$ is *the* appropriate distance measure for a tight upper bound. In fact, it is not – it is merely one of infinitely many appropriate distance measures that all lead to the same upper bound. All we exploited in the proof is that $\sqrt{n-\text{ONEMAX}(x)}$ is concave. One can generalize the proof of Theorem 2 and show the same upper bound using any distance measure $(n-\text{ONEMAX}(x))^\varepsilon$ with any constant $0<\varepsilon<1$ [16].

We now know the expected optimization time of PO-EA, it is $\Theta(n^{3/2})$. This, however, is an asymptotic result. The result itself does not tell us anything about multiplicative constants or terms of smaller order influencing the actual expected optimization time considerably. It does not even tell us whether it is valid for small values of n. We know from the proofs, however, that we do not need to have particularly large values of n to have our arguments be valid. Still it makes sense to consider empirical run times from actual runs to give us at least some intuition about the details that were lost in our asymptotic calculations.

Here, we present results of a straightforward implementation of the PO-EA. We count the number of rounds the algorithm performs before reaching the global optimum 1^n. We present averages over 100 independent runs for each value of n together with their 95% confidence intervals. The averages are plotted as empty circles in Fig. 1, the confidence intervals as small bars. Values of n used are $n\in\{10,20,30,\ldots,4490\}$.

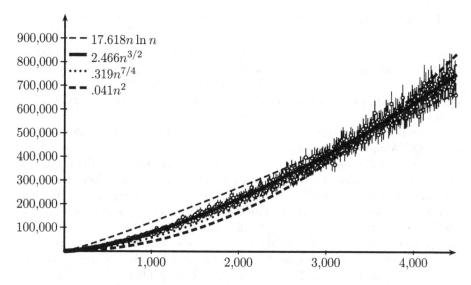

Fig. 1. Results of runs together with fitted curves

In order to obtain a more complete picture we compare this empirical data with $\Theta(n \log n)$, the expected optimization time of the $(1+1)$ EA on a linear function, with $\Theta(n^{3/2})$, the expected optimization time of PO-EA, with $\Theta(n^{7/4})$, the slightly simpler upper bound obtained by using drift analysis with distance measure $\sqrt{n - \text{ONEMAX}(x)}$ alone, and with $\Theta(n^2)$, the upper bound obtained by using drift analysis with the number of 0-bits as distance measure. Since we need to plot actual functions, we decide to plot the functions $c_1 \cdot n \log n$, $c_2 \cdot n^{3/2}$, $c_3 \cdot n^{7/4}$, and $c_4 \cdot n^2$ with some constant factors $c_1, c_2, c_3, c_4 \in \mathbb{R}^+$. To allow for a fair comparison we obtain the fixed coefficient $c_1, \ldots, c_4$ by using gnuplot's fit (gnuplot in version 4.0). We list the actual functions plotted together with RMS (root mean square) of residuals in Table 1.

Table 1. Functions plotted in Fig. 1 together with the RMS of residuals

Function	RMS of Residuals
$17.618n \log n$	42491.6
$2.466n^{3/2}$	18288.6
$.319n^{7/4}$	28370.2
$.041n^2$	45034.1

The results from Table 1 clearly show that $2.466n^{3/2}$ fits the empirical data best. Since the expected optimization time actually is $\Theta(n^{3/2})$, this could be expected. We observe, however, that this information cannot be obtained by heuristic arguments and empirical data, alone. If we only had an upper bound of $O(n^{7/4})$, the empirical data may lead us to believe that this is the actual expected optimization time since the curve of $.319n^{7/4}$ fits the data points reasonably well. This demonstrates the usefulness of theoretical results.

5   Conclusions

Considering the (1+1) EA on linear functions, we devised an abstract model of the fitness landscapes defined by linear functions. The model is pessimistic with respect to the number of 1-bits. Yet it is exact in some situations. In spite of the similarities to linear functions, the expected optimization time is significantly larger.

It is interesting to observe how brittle the (1+1) EA can be. The difference between PO-EA and the (1+1) EA on linear functions seems to be quite marginal. The most probable mutations are mutations of single bits. For these type of mutations, PO-EA does not differ from linear functions at all. Even if we take mutations of at most two bits into account, the expected optimization time for PO-EA is $\Theta(n \log n)$. And yet the mutations of several bits simultaneously are able to increase the running time by a factor of $\Theta(n/\log n)$.

In order to strengthen analytical results on the optimization time it is desirable to see what kind of modifications to the fitness landscape have non-zero yet bounded influence on an EAs performance. This is different from describing modifications of fitness functions that do not have any influence on the optimization time at all [2]. Allowing some real influence on an EAs performance and yet bounding the increase in the expected optimization time introduced by this modification is a more difficult task. Here, one non-trivial example has been presented.

Acknowledgments

The author wants to thank Carsten Witt for very helpful discussions. The author was partly supported by the DFG as part of the Collaborative Research Center "Computational Intelligence" (SFB 531).

References

1. Cormen, T.H., Leiserson, C.E., Rivest, R.L., Stein, C.: Introduction to Algorithms, 2nd edn. MIT Press, Cambridge (2001)
2. Droste, S., Jansen, T., Tinnefeld, K., Wegener, I.: A new framework for the valuation of algorithms for black-box optimization. In: De Jong, K.A., Poli, R., Rowe, J.E. (eds.) Foundations of Genetic Algorithms 7 (FOGA 2002), pp. 253–270. Morgan Kaufmann, San Francisco (2003)
3. Droste, S., Jansen, T., Wegener, I.: A rigorous complexity analysis of the (1+1) evolutionary algorithm for separable functions with Boolean inputs. Evolutionary Computation 6(2), 185–196 (1998)
4. Droste, S., Jansen, T., Wegener, I.: On the analysis of the (1+1) evolutionary algorithm. Theoretical Computer Science 276, 51–81 (2002)
5. Droste, S., Jansen, T., Wegener, I.: Optimization with randomized search heuristics — the (A)NFL theorem, realistic scenarios, and difficult functions. Theoretical Computer Science 287(1), 131–144 (2002)

6. Feller, W.: Introduction to Probability Theory and Its Applications, vol. 2. Wiley, Chichester, UK (1966)
7. He, J., Yao, X.: Drift analysis and average time complexity of evolutionary algorithms. Artificial Intelligence 127(1), 57–85 (2001)
8. He, J., Yao, X.: Erratum to: Drift analysis and average time complexity of evolutionary algorithms: Artificial Intelligence 127, 57–85 (2001) Artificial Intelligence 140(1), 245–248 (2002)
9. He, J., Yao, X.: A study of drift analysis for estimating computation time of evolutionary algorithms. Natural Computing 3(1), 21–35 (2004)
10. Jansen, T., De Jong, K.A., Wegener, I.: On the choice of the offspring population size in evolutionary algorithms. Evolutionary Computation 13(4), 413–440 (2005)
11. Jansen, T., Wegener, I.: On the choice of the mutation probability for the (1+1) EA. In: Deb, K., Rudolph, G., Lutton, E., Merelo, J.J., Schoenauer, M., Schwefel, H.-P., Yao, X. (eds.) Parallel Problem Solving from Nature-PPSN VI. LNCS, vol. 1917, pp. 89–98. Springer, Heidelberg (2000)
12. Jansen, T., Wegener, I.: Evolutionary algorithms — how to cope with plateaus of constant fitness and when to reject strings of the same fitness. IEEE Transactions on Evolutionary Computation 5(6), 589–599 (2002)
13. Mitchell, M.: An Introduction to Genetic Algorithms. MIT Press, Cambridge (1998)
14. Mühlenbein, H.: How genetic algorithms really work: mutation and hillclimbing. In: Männer, R., Manderick, B. (eds.) Parallel Problem Solving from Nature (PPSN II). North-Holland, pp. 15–26 (1992)
15. Storch, T.: On the choice of the population size. In: Deb, K., et al. (eds.) GECCO 2004. LNCS, vol. 3102, pp. 748–760. Springer, Heidelberg (2004)
16. Witt, C.: Personal communication (2007)

Mutative Self-adaptation on the Sharp and Parabolic Ridge

Silja Meyer-Nieberg[1] and Hans-Georg Beyer[2]

[1] Department of Computer Science,
Universität der Bundeswehr München,
85577 Neubiberg, Germany
silja.meyer-nieberg@unibw.de
[2] Department of Computer Science,
Vorarlberg University of Applied Sciences,
Hochschulstr. 1, A-6850 Dornbirn, Austria
hans-georg.beyer@fhv.at

Abstract. In this paper, the behavior of intermediate $(\mu/\mu_I, \lambda)$-ES with self-adaptation is considered for two classes of ridge functions: the sharp and the parabolic ridge. Using a step-by-step approach to describe the system's dynamics, we will investigate the underlying causes for the different behaviors of the ES on these function types and the effects of intermediate recombination.

1 Introduction

Evolution strategies (ES) are population-based search heuristics that move through the search space by means of variation, i.e. mutation and recombination, and selection. Their performance strongly depends on the choice of the mutation strength. During an optimization run, the mutation strength must be continuously adapted to allow the ES to travel with sufficiently speed. To this end, several methods have been developed – e.g. Rechenberg's well-known 1/5th-rule [1] or the cumulative step-size adaptation (CSA) and covariance matrix adaptation (CMA) of Ostermeier, Gawelczyk, and Hansen e.g. [2,3].

In this paper, we will concentrate on the self-adaptation mechanism of the mutation strength introduced by Rechenberg [1] and Schwefel [4]. Here, the adjustment of the mutation strength is left to the ES itself. The mutation strength becomes a part of the individual's genome – undergoing variation and selection processes. The mutation strengths that lead to individuals with high fitness values "survive" and can be passed to the next generation.

Theoretical analyses fall into three main approaches: The first considers the Markov chain that results from the ES's dynamics [5,6]. The second approach tries to answer the question of convergence or divergence and studies induced martingale or super-martingales, respectively [7,8,9,10]. The third [11], applies a step-by-step approach – extracting the important features of the stochastic process and deriving approximate equations. Most of the work focuses on the sphere model, i.e., on functions that in- or decrease monotonically with the distance to the optimizer. Ridge functions can be seen

C.R. Stephens et al. (Eds.): FOGA 2007, LNCS 4436, pp. 70–96, 2007.

as an extension of the sphere. Using an orthogonal representation [12], they comprise functions of the form

$$F(\mathbf{y}) = y_1 + d\Big(\sum_{i=2}^{N} y_i^2\Big)^{\alpha/2} =: x - dR^\alpha \qquad (1)$$

where $x := y_1$ denotes the position on the ridge axis, whereas $R := \|(y_2, \ldots, y_N)^\mathrm{T}\|$ is the $(N-1)$-dimensional distance to the ridge. Ridge functions contain a sphere model weighted with parameter d and a linear component. On the first sight, progress is possible by minimizing the sphere and by maximizing the linear part which might be problematic when using isotropic mutations.

The performance of evolution strategies on this function class was already addressed e.g. in [13,12] without considering self-adaptation. The first theoretical analysis of adaptive ES on the ridge function class considering the CSA-ES on the sharp ($\alpha = 1$) and parabolic ridge ($\alpha = 2$) was sketched in [14]. It was found that ridge functions resemble a noisy sphere model which allows for an application of the theory developed there. In [15] a CSA-ES on the noisy parabolic ridge was considered. Three types of noise were analyzed which differed in the increase of the noise strength as a function of the distance to the ridge axis.

In this paper, we will make use of the approach introduced in [11] and consider the self-adaptation behavior of ES on the ridge function class – concentrating on the sharp and parabolic ridge. The ES considered here are intermediate $(\mu/\mu_I, \lambda)$-ES with self-adaptation of a single mutation strength. The offspring are thus generated according to:

1. Compute the mean $\langle\sigma\rangle = \frac{1}{\mu}\sum_{m=1}^{\mu}\sigma_m$ of the mutation strengths of the μ individuals of the parent population
2. Compute the centroid $\langle\mathbf{y}\rangle = \frac{1}{\mu}\sum_{m=1}^{\mu}\mathbf{y}_m$ of the object vectors of the μ individuals of the parent population
3. For all offspring $l \in \{1, \ldots, \lambda\}$:
   (a) To derive the new mutation strength: Mutate the mean $\langle\sigma\rangle$ according to $\sigma_l = \langle\sigma\rangle\zeta$ where ζ is a random variable which should fulfill $\mathrm{E}[\zeta] \approx 1$ (see [16] for a discussion of this and further requirements). Typical choices of ζ's distribution include the log-normal distribution, derivatives of normal distributions, or a two-point distribution [17].
   (b) Generate the object vector $\mathbf{y}_l$ according to $y_i = \langle y_i\rangle + \sigma_l\mathcal{N}(0,1)$ where y_i is the vector's ith component and $\mathcal{N}(0,1)$ stands for a standard normally distributed random variable.

Afterwards, the μ best offspring are chosen – according to their fitness. They (along with their mutation strengths) become the parents of the next generation. We will consider the log-normal operator to mutate the mutation strength. Therefore, the random variable ζ is given by $\zeta = e^{\tau\mathcal{N}(0,1)}$ where τ is called the learning parameter.

This paper is structured as follows. First, we will describe how the ES's dynamics may be modeled. Afterwards, we will introduce the progress measures required before analyzing the behavior of self-adaptive intermediate ES on the sharp and parabolic ridge.

1.1   Modeling the Evolutionary Dynamics: The Evolution Equations

Due to the form of the fitness function which can be given as $f(x, R) = x - dR^{\alpha}$, three variables are of interest. The first is the x-component denoting the change on the ridge axis. The second is the component R measuring the distance to the ridge. The third is the mutation strength ς. Their one-generational change is described by difference equations (called evolution equations)

$$
\begin{aligned}
x^{(g+1)} &= x^{(g)} + \mathrm{E}[x^{(g+1)} - x^{(g)} | (R^{(g)}, x^{(g)}, \langle \varsigma^{(g)} \rangle)] + \mathcal{R}_x^{(g)} \\
&=: x^{(g)} + \varphi_x(R^{(g)}, x^{(g)}, \langle \varsigma^{(g)} \rangle) + \mathcal{R}_x^{(g)}
\end{aligned}
\tag{2}
$$

$$
\begin{aligned}
R^{(g+1)} &= R^{(g)} + \mathrm{E}[R^{(g+1)} - R^{(g)} | (R^{(g)}, x^{(g)}, \langle \varsigma^{(g)} \rangle)] + \mathcal{R}_R^{(g)} \\
&=: R^{(g)} - \varphi_R(R^{(g)}, x^{(g)}, \langle \varsigma^{(g)} \rangle) + \mathcal{R}_R^{(g)}
\end{aligned}
\tag{3}
$$

$$
\begin{aligned}
\langle \varsigma^{(g+1)} \rangle &= \langle \varsigma^{(g)} \rangle \left(1 + \mathrm{E}\left[\frac{\langle \varsigma^{(g+1)} \rangle - \langle \varsigma^{(g)} \rangle}{\langle \varsigma^{(g)} \rangle} | (R^{(g)}, x^{(g)}, \langle \varsigma^{(g)} \rangle) \right] \right) + \mathcal{R}_{\varsigma}^{(g)} \\
&=: \langle \varsigma^{(g)} \rangle \left(1 + \psi(R^{(g)}, x^{(g)}, \langle \varsigma^{(g)} \rangle) \right) + \mathcal{R}_{\varsigma}^{(g)}
\end{aligned}
\tag{4}
$$

which consist of a deterministic part, i.e., the (conditional) expected change and a perturbation part $\mathcal{R}$ covering the random fluctuations. Note, in the case of the mutation strength a multiplicative change is considered. In a first analysis, we will neglect the perturbation parts of the evolution equations. Therefore, it remains to determine the conditional expected changes of the state variables. In the case of the x- and R-variables, these are the progress rates φ_x and φ_R. In the case of the mutation strength, we have to determine the so-called self-adaptation response (SAR) ψ denoting the expected relative change. To simplify the notations, we will set $\sigma := \langle \varsigma^{(g)} \rangle$, $R := R^{(g)}$, $r := R^{(g+1)}$, $x := x^{(g+1)}$, and $X := x^{(g)}$ unless the dependence on the generation number g is explicitly needed. Similarly, we will not denote the dependency of the expected changes, i.e., the progress rates and the SAR, on the previous values.

2   Preliminaries

In this section, we will give a short sketch of the derivation of the progress measures required. More detailed versions can be found in the appendices.

In the case of the progress rates, we follow a similar way as in [15] and [12]. A main point concerns the derivation of a probability density function (pdf) which describes the fitness or quality change by a mutation. The new vector of an offspring consists of a component y_1 (i.e., along the axis) and of a perpendicular component $\mathbf{r} = (0, y_2, \dots, y_N)^{\mathrm{T}}$. Its length $r = \|\mathbf{r}\|$ denotes the distance to the ridge. As in the case of the sphere model, this $(N-1)$-dimensional $\mathbf{r}$ can be decomposed into $\mathbf{r} = \mathbf{R} - z_R \mathbf{e}_R + \mathbf{h}$, with z_R the component of the mutation vector in $-\mathbf{R}$-direction and $\mathbf{h}$ the part perpendicular to $\mathbf{R}$. The second important point is the linearization of the distance r in its components which allows further treatment in the case of the sharp ridge. The feasibility of the approach depends on the assumption that the change of the distance in one generation is small (see [12]).

All progress measures –the progress rates as well as the SAR– give the change of the centroid or mean, respectively in the case of the SAR, of the state variables x, R, and ς of the μ best offspring. The offspring l are chosen according to their fitness values $F(\mathbf{y}_l)$ or to the fitness (quality) change $Q_l := F(\mathbf{y}_l) - F(\langle \mathbf{y} \rangle)$ which is more convenient for the calculations. Ordering the λ offspring after non-decreasing values of Q_l, we obtain the usual ordering $Q_{1:\lambda} \leq \ldots \leq Q_{\lambda:\lambda}$ of order statistics [18], where $Q_{m:\lambda}$ denotes the mth smallest of λ trials. Note, for the progress measures the quality change is not used directly but we need the distributions of the first component x, distance r, and mutation strength ς which are associated with the mth highest quality change. Therefore, the so-called induced order statistics is applied [19].

The progress rate φ_x describing the expected one-generation change in ridge direction is defined as

$$\varphi_x := \mathrm{E}[x^{(g+1)} - x^{(g)}]. \tag{5}$$

It gives the expected change of the first component of the centroid vector. Using Lemma 1 of Appendix B the progress rate can be easily determined (see p. 88). Using the normalization $\varphi_x^* = N\varphi_x$ and letting $N \to \infty$

$$\varphi_x^*(\sigma^*, R) = \frac{\sigma^*}{\sqrt{1+q^2}} c_{\mu/\mu,\lambda} \tag{6}$$

with

$$q := d\alpha R^{\alpha-1} \tag{7}$$

is obtained. The coefficient $c_{\mu/\mu,\lambda}$ in (6) is a special case of the generalized progress coefficients [20, p.172]

$$e_{\mu,\lambda}^{\alpha,\beta} = (\lambda - \mu) \binom{\lambda}{\mu} \frac{1}{\sqrt{2\pi}^{\alpha+1}} \int_{-\infty}^{\infty} t^\beta e^{-\frac{(\alpha+1)t^2}{2}} \Phi(t)^{\lambda-\mu-1} (1 - \Phi(t))^{\mu-\alpha} \, dt \tag{8}$$

with $c_{\mu/\mu,\lambda} := e_{\mu,\lambda}^{1,0}$. We will restrict the analysis to $N \gg 1$. Therefore $N - 1 \approx N$, and as in (6) the normalizations $\varsigma^* = N\varsigma$ and $\varphi_x^* = N\varphi_x$ are used.

The progress rate φ_x^* (6) has no loss term and increases linearly with the mutation strength. Thus, the expected progress on the ridge axis is always positive but goes to zero for $\varsigma^* \to 0$ or $q \to \infty$. As can be seen, the progress rate is influenced by the ridge function itself, more specifically, it is influenced by the gradient vector

$$\nabla f(x, R) = \begin{pmatrix} 1 \\ -\alpha dR^{\alpha-1} \end{pmatrix}. \tag{9}$$

The expression $1/\sqrt{1 + (\alpha dR^{\alpha-1})^2}$ which appears in the progress rate (6) equals the cosine of the angle between the gradient and the ridge axis. With the exception of the sharp ridge where the angle is constant, the farer the ES is away from the ridge the steeper the slope of gradient, the higher the angle and the smaller the cosine and with it the progress rate parallel to the axis.

The progress rate of the R-evolution is defined by

$$\varphi_R := \mathrm{E}[R - r] \tag{10}$$

where r denotes the length of the $(N - 1)$-dimensional centroid vector of the μ best offspring or the next parental population, respectively. As shown in [20] and used in Appendix B, the new vector $\mathbf{r}$ can be written as the sum of two orthogonal vectors, the first denoting the change with respect to the old centroid vector $\mathbf{R}$, the other perpendicular to that. The new vector can thus be given by $\mathbf{r} = \mathbf{R} - \langle z_R \rangle \mathbf{e_R} + \langle \mathbf{h} \rangle$, where $\langle \mathbf{h} \rangle$ denotes the component perpendicular to $\mathbf{R}$ and $\mathbf{e_R} := \mathbf{R}/R$. Since $r = \|\mathbf{r}\| = \sqrt{\mathbf{r}^\mathrm{T}\mathbf{r}}$, the progress rate (10) can also be written as

$$\varphi_R := \mathrm{E}[R - \sqrt{(R - \langle z_R \rangle)^2 + \langle h \rangle^2}]. \tag{11}$$

As shown in Appendix B, this decomposition can be used to derive

$$\varphi_R^*(\sigma^*, R) = \frac{q\sigma^*}{\sqrt{1 + q^2}} c_{\mu/\mu,\lambda} - \frac{\sigma^{*2}}{2R\mu} \tag{12}$$

for $N \to \infty$ which will be used as an approximate equation for finite N. The parameter q is denotes again $q := d\alpha R^{\alpha-1}$, (7). The progress rate is similar to that on the sphere model [20] – with a loss part which is a result of the perpendicular component of the $\mathbf{r} = (y_2, \ldots, y_N)^\mathrm{T}$-vector (see Appendix B). The ridge factor enters the equation over the q-values (7) and influences the gain but not the loss part of the progress rate. The influence is similar to the progress rate of the x-evolution. This time the influence of the gradient (9) at R and $x^{(g)}$ enters over the sine of the angle between the vector and the x-axis. For increasing distances to the ridge, the sine approaches one and the angle $\pi/2$. More and more weight is put on the gain part, i.e. on the part that stems from the mutation vector components pointing in $\mathbf{R}$-direction. For decreasing distances, the sine and the angle go to zero and the gain part looses influence. The exception is again the sharp ridge with a constant angle.

Both progress rates (6) and (12) are obtained for the case of $\tau = 0$ and therefore only applicable for small values of τ.

The self-adaptation response (SAR) denotes the expected relative change of the mutation strength during one generation

$$\psi(\sigma^*, X, R) = \frac{1}{\mu} \sum_{m=1}^{\mu} \mathrm{E}\left[\frac{\varsigma_{m;\lambda}^* - \sigma^*}{\sigma^*} \right]$$

$$= \int_0^\infty \left(\frac{\varsigma^* - \sigma^*}{\sigma^*} \right) p_{m;\lambda}(\varsigma^*, \sigma^*, X, R) \, \mathrm{d}\varsigma^* \tag{13}$$

where $p_{m;\lambda}(\varsigma^*)$ denotes the density function of the mutation strength of the mth best offspring. It should be noted that a closed analytical derivation of the SAR does not appear feasible. Therefore several simplifications have to be introduced. One of the most important points is to restrict the analysis to $\tau \ll 1$ provided that the log-normal operator is used. The learning parameter is generally chosen to scale with $1/\sqrt{N}$. Provided that the search space dimensionality is sufficiently high, this should not be a severe restriction. The reason for requiring $\tau \ll 1$ is due to the finding that it is possible to simplify several integral expressions introducing an error of order $\mathcal{O}(\tau^2)$ which can be assumed to be negligible provided τ is sufficiently small (note, due to further integrations the error term will finally be of order $\mathcal{O}(\tau^4)$). The derivation of the SAR follows the approach taken in [20]. The first point is to find an expression for the probability density function (pdf) of the mth best offspring. Afterwards it remains to solve the resulting integrals and to cope with the sum in (13) (see Appendix C.1). We obtain for $N \to \infty$ the SAR

$$\psi_\infty(\sigma^*) = \tau^2 \left(\frac{1}{2} + e^{1,1}_{\mu,\lambda} - c_{\mu/\mu,\lambda} \frac{\sigma^*}{R} \frac{q}{\sqrt{1+q^2}} \right) + \mathcal{O}(\tau^4) \qquad (14)$$

with $q := d\alpha R^{\alpha-1}$ (7). The SAR has a linear loss part – again influenced by the distance to the ridge over the fraction and the sine of the slope angle of the gradient – and a constant positive component. Figure 1 compares (14) with the results of experiments for $N = 30$. Some deviations can be observed – especially for larger σ^*, but generally the prediction quality appears to be sufficient.

Equations (6), (12), and (14) describe the expected changes of the state variables under the conditions of $\tau \ll 1$ and $N \to \infty$. As mentioned, they serve as approximate formulae in finite dimensional search spaces. So, now we are in the position to analyze the expected changes on the one hand, i.e., under which conditions a positive or negative change is expected, and on the other hand to take a closer look at the ES's evolution if the random fluctuations are neglected.

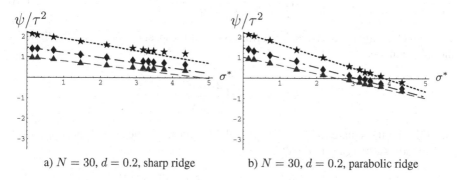

a) $N = 30$, $d = 0.2$, sharp ridge   b) $N = 30$, $d = 0.2$, parabolic ridge

Fig. 1. The first-order SAR (Eq. (14)) on the sharp and parabolic ridge for some $(\mu/\mu, 10)$-ES. Shown are the results for $\mu = 1$ (black, stars), $\mu = 2$ (blue, diamonds), and $\mu = 3$ (red, triangles). The distance to the ridge was set to $R = 1$. Each data point was obtained by sampling over $100,000$ one-generation experiments.

3   Self-adaptation on the Sharp and Parabolic Ridge

As observed in numerical experiments, self-adaptive ES may experience problems on ridge functions. Especially on the sharp ridge, they are known to converge prematurely – optimizing only the sphere-part of the model (see e.g. [21]). Let us consider the evolution equations (with $\varsigma^* = \varsigma N$, $\varphi^*_R = N\varphi_R$, and $\varphi^*_x = N\varphi_x$)

$$x^{(g+1)} = x^{(g)} + \frac{1}{N}\varphi_x^*(R^{(g)}, \langle \varsigma^{*(g)} \rangle)$$

$$R^{(g+1)} = R^{(g)} - \frac{1}{N}\varphi_R^*(R^{(g)}, \langle \varsigma^{*(g)} \rangle)$$

$$\langle \varsigma^{*(g+1)} \rangle = \langle \varsigma^{*(g)} \rangle(1 + \psi(R^{(g)}, \langle \varsigma^{*(g)} \rangle)) \tag{15}$$

with the SAR (14) and progress rates (6) and (12) obtained for $N \to \infty$. Note, this is only a very rough approximation of real systems with small N and small population sizes. The results obtained by analyzing (15) must therefore be compared with experiments. We will start by considering the system in R and ς^*. Note, according to (15) there is no feedback of the x-evolution on these two state variables.

3.1 The System in ς^* and R

Let us start with the evolution of the mutation strength. The present mutation strength is increased if the value of the SAR (14) is positive and decreased otherwise. The SAR is a monotonously decreasing function in ς^* with only one zero $\varsigma^*_{\psi_0}$ which depends on the ridge factor $q = d\alpha R^{\alpha-1}$ (7) and $R = R^{(g)}$

$$\varsigma^*_{\psi_0} = R\frac{1/2 + e^{1,1}_{\mu,\lambda}}{c_{\mu/\mu,\lambda}}\sqrt{\frac{1 + \alpha^2 d^2 R^{2\alpha-2}}{\alpha^2 d^2 R^{2\alpha-2}}}$$

$$= R\varsigma^{*sph}_{\psi_0}\sqrt{\frac{1 + \alpha^2 d^2 R^{2\alpha-2}}{\alpha^2 d^2 R^{2\alpha-2}}}. \tag{16}$$

The zero (16) only differs from the normalized (with respect to N) zero of the SAR for the sphere model

$$\varsigma^{*sph}_{\psi_0} := R\frac{1/2 + e^{1,1}_{\mu,\lambda}}{c_{\mu/\mu,\lambda}}. \tag{17}$$

(see [22]) by the square root which equals the reciprocal of the sine of the slope angle of the gradient. It is easy to see that

$$\lim_{R\to\infty}\varsigma^*_{\psi_0} = \frac{1/2 + e^{1,1}_{\mu,\lambda}}{c_{\mu/\mu,\lambda}}\lim_{R\to\infty}\sqrt{\frac{1 + \alpha^2 d^2 R^{2\alpha-2}}{\alpha^2 d^2 R^{2\alpha-4}}} = \infty$$

$$\lim_{R\to 0}\varsigma^*_{\psi_0} = \begin{cases} \infty & \text{if } \alpha > 2 \\ \frac{1/2+e^{1,1}_{\mu,\lambda}}{2dc_{\mu/\mu,\lambda}} & \text{if } \alpha = 2 \\ 0 & \text{if } \alpha = 1 \end{cases} \tag{18}$$

holds. As one can see, if R increases, the SAR tends to increase higher and higher mutation strengths in turn. On the other hand, for decreasing distances, the SAR also answers with increasing higher and higher mutation strengths ($\alpha > 2$) or mutation strengths higher than a limit value ($\alpha = 2$). The important point is that the ES maintains a strictly positive mutation strength – provided that $\alpha > 1$. In the case of the sharp ridge it goes to zero. These behaviors can be traced back to the shape of the ridge or more correctly to the gradient at $R \nabla f(x, R) = (1, -d\alpha R^{\alpha-1})^{\mathsf{T}}$. Let us reconsider the SAR

$$\psi(\sigma^*) = \tau^2 \left(1/2 + e_{\mu,\lambda}^{1,1} - \frac{d\alpha R^{\alpha-1}}{\sqrt{1 + (d\alpha R^{\alpha-1})^2}} \frac{\sigma^*}{R} \right).$$

In the case of the sharp ridge, the slope of the gradient and with it the sine of the angle is constant through the search space. As the result, the SAR behaves as in the case of the sphere model. As the distance decreases, smaller and smaller mutation strengths lead to a negative SAR and to an expected decrease of the mutation strength. In the case of the sphere model this behavior is desirable and necessary: If the ES approaches the optimum, the mutation strength should be decreased. In the case of ridge functions this indicates a potential premature convergence.

The slope of the gradient of quadratic (or higher) ridge functions depends in stark contrast to the sharp ridge on the position in the search space or more correctly on how far the ES is away from the ridge. As the distance to the axis decreases, the angle between axis and gradient becomes smaller. The sine approaches zero and counteracts to some extend the normal reaction from the sphere model to increase the loss part.

The R-evolution remains to be considered. The progress rate φ_R^* (12) is strictly positive in the interval $\varsigma^* \in (0, 2R\mu\, c_{\mu/\mu,\lambda}\, \sqrt{q^2/(1+q^2)})$. The second zero of the progress rate reads

$$\varsigma_{\varphi R}^* = 2R\mu c_{\mu/\mu,\lambda} \sqrt{\frac{\alpha^2 d^2 R^{2\alpha-2}}{1 + \alpha^2 d^2 R^{2\alpha-2}}}$$

$$= R\varsigma_{\varphi R}^{*\, sph} \sqrt{\frac{\alpha^2 d^2 R^{2\alpha-2}}{1 + \alpha^2 d^2 R^{2\alpha-2}}} \qquad (19)$$

with

$$\varsigma_{\varphi R}^{*\, sph} := R2\mu c_{\mu/\mu,\lambda} \qquad (20)$$

denoting the normalized (with respect to N) zero of the progress rate in the case of the sphere model [20]. Again the zero of the sphere model appears which is weighted in this case by the sine of the slope angle of the gradient and not by its reciprocal as in the case of the zero of the SAR. As a result, it can be shown that the zero of the progress rate behaves in accordance with the distance to the ridge, i.e.,

$$\lim_{R \to \infty} \varsigma_{\varphi R}^* = 2\mu c_{\mu/\mu,\lambda} \lim_{R \to \infty} \sqrt{\frac{\alpha^2 d^2 R^{2\alpha}}{1 + \alpha^2 d^2 R^{2\alpha-2}}} = \infty$$

$$\lim_{R \to 0} \varsigma_{\varphi R}^* = 2\mu c_{\mu/\mu,\lambda} \lim_{R \to 0} \sqrt{\frac{\alpha^2 d^2 R^{2\alpha}}{1 + \alpha^2 d^2 R^{2\alpha-2}}} = 0. \qquad (21)$$

As we have seen, one of the first obvious differences between the sharp ridge ($\alpha = 1$) and higher order ridge functions ($\alpha \geq 2$) is that only in the case of the latter, the SAR (14) does not allow the mutation strength to mirror a decrease of the distance infinitely. Furthermore, only for $\alpha = 1$ both zeros (16) and (19) are linear functions in R. To discuss these and further differences, we will now take a closer look at the sharp and parabolic ridge.

3.2 The Parabolic Ridge

Let us first consider the parabolic ridge, i.e. $\alpha = 2$ and $F(\mathbf{y}) = x - dR^2$, as a representative of ridge functions with $\alpha \geq 2$. Figure 2 illustrates the behavior of the (ς^*, R)-system depicting the so-called isoclines (see e.g. [23]) $\varphi_R^* = 0$ and $\psi = 0$ as functions of R. The area below the isocline $\psi = 0$ is characterized by (ς^*, R)-combinations for which the SAR is positive and the mutation strength is expected to increase. Similarly, the area below $\varphi_R^* = 0$ is characterized by a positive progress rate and thus by an expected decrease of the distance to the ridge. If R increases, the SAR tends to increase higher and higher mutation strengths. This effects in turn the R-evolution. Here, the zero of the progress rate increases as well. Higher and higher mutation strengths will result in an expected decrease of the distance and not in a further increase. On the other hand, if R decreases, the zero of the progress rate decreases as well. Mutation strengths that would increase the distance are thus also decreasing. But the potential answer of the ς-evolution is either to increase an increasing range of mutation strengths or at least every mutation strength smaller than a limit. Thus, neither a convergence of $R \to 0$, i.e., a convergence to the axis, nor a divergence of $R \to \infty$ occurs.

As Fig. 2 shows there is a stationary state of the (ς^*, R)-system. In the stationary state the ς^*- and the R-evolution come to hold (on average) – i.e., the evolution strategy is expected to fluctuate at a certain distance to the axis. Apart from the trivial stationary state with $\varsigma^* = 0$, the stationary state is characterized by requiring that both the SAR (14) and the progress rate φ_R^* (12) are zero.

Considering the last two equations of (15), their stationary points are given by

$$R^{(g+1)} = R^{(g)} \Rightarrow \varphi_R(R^{(g)}, \langle \varsigma^{*(g)} \rangle) = 0$$
$$\langle \varsigma^{*(g+1)} \rangle = \langle \varsigma^{*(g)} \rangle \Rightarrow \psi(R^{(g)}, \langle \varsigma^{*(g)} \rangle) = 0. \tag{22}$$

As mentioned, the progress rate (12) and the SAR (14) must be zero. In order to get the stationary R, one has to solve both equations simultaneously for R. Taking (16) and (19) into account we obtain by $\varsigma^*_{\psi_0} = \varsigma^*_{\varphi_R}$, i.e.,

$$R\left(\frac{1/2 + e_{\mu,\lambda}^{1,1}}{c_{\mu/\mu,\lambda}}\right)\sqrt{\frac{1+q^2}{q^2}} = 2\mu c_{\mu/\mu,\lambda} R \sqrt{\frac{q^2}{1+q^2}}$$

$$\Rightarrow q_{st}^2 = \frac{1/2 + e_{\mu,\lambda}^{1,1}}{2\mu c_{\mu/\mu,\lambda}^2 - 1/2 - e_{\mu,\lambda}^{1,1}}. \tag{23}$$

Since $q^2 = \alpha^2 d^2 R^{2\alpha-2}$, (7), the stationary distance to the ridge becomes

$$R_{st} = \left(\frac{1/2 + e_{\mu,\lambda}^{1,1}}{\alpha^2 d^2 \left(2\mu c_{\mu/\mu,\lambda}^2 - 1/2 - e_{\mu,\lambda}^{1,1}\right)}\right)^{1/(2\alpha-2)} \tag{24}$$

for general $\alpha > 1$ and

$$R_{st} = \frac{1}{2d}\sqrt{\frac{1/2 + e_{\mu,\lambda}^{1,1}}{2\mu c_{\mu/\mu,\lambda}^2 - 1/2 - e_{\mu,\lambda}^{1,1}}} \tag{25}$$

for the parabolic ridge itself. As one can see, it decreases with increasing d.

In [14] an estimate of the stationary distance of CSA-ES was obtained for the parabolic ridge, i.e., for $\alpha = 2$, $R_{st} \propto 1/(2d)$ which also reappears in the case of σ-self-adaptation. Again, a similarity with the situation of the noisy sphere model appears [24]. In that case, the stationary distance scales with the standard deviation (noise strength) of additive normally distributed noise, i.e. $R_{st} \propto \sigma_\epsilon$. Therefore, $1/d$, the inversion of the weighting constant of the embedded sphere, appears similar to the noise term σ_ϵ.

The stationary distance (24) can be used together with the SAR's zero (16) or with the zero of the progress rate (19) to give the stationary mutation strength as

$$\varsigma^*_{st} = \sqrt{2\mu(1/2 + e^{1,1}_{\mu,\lambda})} \left(\frac{1/2 + e^{1,1}_{\mu,\lambda}}{\alpha^2 d^2 \left(2\mu c^2_{\mu/\mu,\lambda} - 1/2 - e^{1,1}_{\mu,\lambda}\right)} \right)^{1/(2\alpha - 2)} \tag{26}$$

in the general case and

$$\varsigma^*_{st} = \frac{1/2 + e^{1,1}_{\mu,\lambda}}{2d} \sqrt{\frac{2\mu}{2\mu c^2_{\mu/\mu,\lambda} - 1/2 - e^{1,1}_{\mu,\lambda}}} \tag{27}$$

for $\alpha = 2$. For the progress in direction of the ridge axis, this leads to

$$\varphi^*_{x\,st} = \frac{c_{\mu/\mu,\lambda}}{\sqrt{1 + q^2_{st}}} \varsigma^*_{st}$$

$$= \sqrt{\left(1/2 + e^{1,1}_{\mu,\lambda}\right)\left(2\mu c^2_{\mu/\mu,\lambda} - 1/2 - e^{1,1}_{\mu,\lambda}\right)}$$

$$\times \left(\frac{1/2 + e^{1,1}_{\mu,\lambda}}{\alpha^2 d^2 \left(2\mu c^2_{\mu/\mu,\lambda} - 1/2 - e^{1,1}_{\mu,\lambda}\right)} \right)^{1/(2\alpha - 2)} \tag{28}$$

as the expected stationary progress on the ridge for $\alpha > 1$ leading for $\alpha = 2$ to

$$\varphi^*_{x\,st} = \frac{1/2 + e^{1,1}_{\mu,\lambda}}{2d}.$$

Figure 3 shows the stationary mutation strength (27), distance (25), and progress rate (29) as functions of the parent number μ comparing them with the results of experiments. The agreement between prediction and experiment is good, but it should be mentioned that the experimental data are lower than predicted. Interestingly, the experiments do not show significant differences between high and low search space dimensionality. This is somewhat surprising and up to now not fully understood.

Equations (27) and (29) can be used to investigate the influence of recombination of the object parameters on the self-adaptation response function. As Fig. 3 shows, the maximal progress and the maximal mutation strength occur in the case of non-recombinative ES, i.e., for $\mu = 1$, which is confirmed by experiments. Introducing multi-parent recombination does not lead to any advantage at all.

The stationary progress on the axis is influenced by the stationary mutation strength and therefore by progress rate (towards the axis) (12) and the SAR (14). In the case of the parabolic ridge, it can be given as $\varphi^*_{x\,st} = (1/2 + e^{1,1}_{\mu,\lambda})/(2d)$ (29). The effects of recombination are reflected by the progress coefficient $e^{1,1}_{\mu,\lambda}$ which stems from the SAR. All other influences have averaged out. The progress coefficient $e_{\mu,\lambda}^{1,1}$ is a monotonously decreasing function in μ for $\mu < \lambda/2$. The first zero point is at $\mu = \lambda/2$. Afterwards, negative values are assumed until it approaches zero again for $\mu = \lambda$. As a result, the stationary progress (29) is highest for $\mu = 1$ and ES does not benefit from recombination.

To sum up the findings before addressing the sharp ridge: Using the deterministic variant of the evolution equations, two main results can be derived. First, a stationary state other than $\varsigma^* = 0$ exists which allows for positive progress. Second, the ES does not benefit at all from multi-parent recombination.

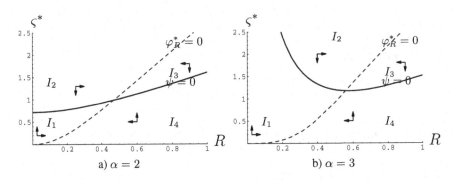

a) $\alpha = 2$ b) $\alpha = 3$

Fig. 2. The zero points of the progress rate φ_R^* and ψ as functions of the distance to the ridge R for $(1, 10)$-ES with $d = 1$. Region I_1 is characterized by $\Delta R > 0$, $\Delta\varsigma^* > 0$, I_2 by $\Delta R < 0$, $\Delta\varsigma^* > 0$, I_3 by $\Delta R < 0$, $\Delta\varsigma^* < 0$, and finally I_4 by $\Delta R > 0$, $\Delta\varsigma^* < 0$. The system either leaves every region I_k a gain, i.e., it oscillates, or it converges to the equilibrium point.

3.3 The Sharp Ridge

For the sharp ridge, i.e., for $\alpha = 1$ and $F(\mathbf{y}) = x - dR$, the ES shows a completely different behavior without a stationary distance to the ridge. In the case of CSA-ES, it was found [14] that the behavior is determined by the choice of the d-parameter of the ridge: Either a convergence towards the axis or a divergence $R \to \infty$ occurs. In contrast to $\alpha > 1$, there is no additional feedback from the distance R over the $q = d\alpha R^{\alpha-1}$ parameter. The stationary state with $\psi = \varphi_R^* = 0$, can only be fulfilled for one choice of the parameter, i.e., for

$$d_{crit} = \sqrt{\frac{1/2 + e_{\mu,\lambda}^{1,1}}{2\mu c_{\mu/\mu,\lambda}^2 - 1/2 - e_{\mu,\lambda}^{1,1}}} \qquad (29)$$

(see (23)). Otherwise, there is no stationary point (other then $R = 0$ or $\varsigma^* = 0$). The critical d-value (29) decides over the question of convergence and divergence. It depends strongly on the population parameter μ and is highest for $\mu = 1$ or μ approaching λ as Fig. 4 illustrates. Recombination lowers the critical d-value for the usual $\mu : \lambda$-ratios. As can be seen for $\mu = 1$ only the results for $d = 0.9$ (critical d-value 0.936) converge. In the case of $\mu = 3$ with a critical d-value of 0.418, all runs with $d < 0.5$ diverge

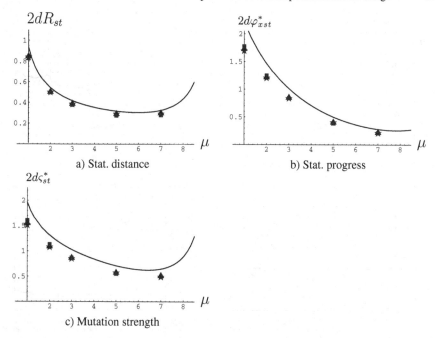

Fig. 3. The stationary distance (25), mutation strength (27), and progress rate (29) for some $(\mu/\mu_I, 10)$-ES with self-adaptation on the parabolic ridge. Each data point was sampled over $200,000$ generations ($N = 30$, $N = 100$) and $900,000$ ($N = 1000$) generations. The stars indicate the results for $N = 1000$, the triangles those for $N = 100$, and the boxes those for $N = 30$.

whereas for $\mu = 5$ only the runs for $d = 0.2$ diverge clearly. The critical d-value in this case is 0.31.

The general situation of the system in ς^* and R is depicted in Fig. 5 which again shows the isoclines $\varphi_R^* = 0$ and $\psi = 0$. Both are now linear functions in R. Again, the area below $\psi = 0$ is characterized by a positive SAR and an expected increase of the mutation strength and the area below $\varphi_R^* = 0$ is characterized by positive progress and an expected decrease of the distance (which is indicated by the arrows in Fig. 5). As can be seen in the figure, the ES will eventually move into the area between the two isoclines. Let us illustrate that by example for Fig. 5 a). Here, the isocline $\varphi_R^* = 0$ is above the isocline $\psi = 0$. This equals the condition $d > d_{crit}$. If the system starts in the area below $\psi = 0$, the SAR and the progress rate are positive. As a result, the mutation strength increases and the distance decreases. The system thus moves towards the line $\psi = 0$. Once this is reached, the ς^*-evolution temporarily stops. But since the R-evolution still progresses and the distance decreases, the isocline $\psi = 0$ is crossed and the system enters the area between both isoclines. There it remains and approaches zero. Therefore for $\varsigma^*_{\varphi_R} > \varsigma^*_{\psi_0}$, i.e. $d > d_{crit}$, the system in R and ς^* approaches the origin with $R \to 0$, $\varsigma^* \to 0$ as in the case of the sphere. On the other hand, once $d < d_{crit}$ (see Fig. 5 b)) and $\varsigma^*_{\varphi_R} < \varsigma^*_{\psi_0}$, the system reaches the cone defined by the $\varphi_R^* = 0$ and $\psi = 0$ but it moves into the opposite direction – going to infinity. It is easy to see that the latter case is always connected with a positive (expected) quality change.

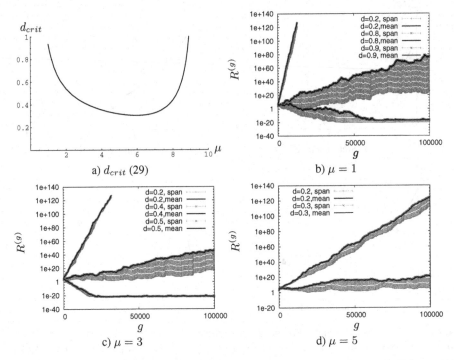

a) d_{crit} (29)   b) $\mu = 1$

c) $\mu = 3$ d) $\mu = 5$

Fig. 4. Results from $(\mu/\mu_I, \lambda)$-ES runs for the first $100,000$ generations for several choices of d ($N = 100$). Shown is every 50th value. Each data line is averaged over 20 runs. Also shown are the minimal and maximal values.

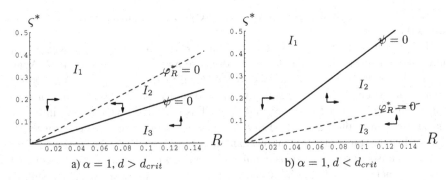

a) $\alpha = 1, d > d_{crit}$ b) $\alpha = 1, d < d_{crit}$

Fig. 5. The isoclines $\varphi_R^* = 0$ and $\psi = 0$ as functions of the distance to the ridge R for $(1, 10)$-ES with $a = 1$. In a) region I_1 is characterized by $\Delta R > 0$, $\Delta\varsigma^* < 0$, I_2 by $\Delta R < 0$, $\Delta\varsigma^* < 0$, and I_3 by $\Delta R < 0$, $\Delta\varsigma^* > 0$. Possible movements between the regions are $I_3 \to I_2$ and $I_1 \to I_2$. It is easy to see that I_1 and I_3 will be left eventually. The region I_2 cannot be left and the system in ς^* and R approaches the origin. In b) region I_1 is characterized by $\Delta R > 0$, $\Delta\varsigma^* < 0$, I_2 by $\Delta R > 0$, $\Delta\varsigma^* > 0$, and I_3 by $\Delta R < 0$, $\Delta\varsigma^* > 0$. Possible movements are $I_1 \to I_2$ and $I_3 \to I_2$, but I_2 cannot be left. The system diverges to infinity.

So, if λ and μ are chosen appropriately to d, the self-adaptive ES does not converge prematurely but shows the behavior required.

The question that remains, though, is whether the ES is able to travel with nearly optimal speed. Let $d < d_{crit}$ and consider the quality change $\Delta Q^* = \varphi_x^* + d\varphi_R^*$

$$\Delta Q^* = \sqrt{1+d^2}c_{\mu/\mu,\lambda}\varsigma^* - \frac{d}{2R\mu}\varsigma^{*2} \tag{30}$$

as the performance measure. Its optimizer is given by $\varsigma_{opt}^* = c_{\mu/\mu,\lambda}R\mu\sqrt{1+d^2}/d$. As Fig. 5 shows the ES – i.e., the system in ς^* and R is expected to remain in region I_2 in the long run. The maximal mutation strength the ES can attain there is the SAR's zero $\varsigma_\psi^* = R(1/2 + e_{\mu,\lambda}^{1,1})/c_{\mu/\mu,\lambda}\sqrt{1+d^2}/d$ (see (16)). It can be shown by case inspection that for a long range of μ-values (except for $\mu = 1$) ς_ψ^* is quite smaller than ς_{opt}^* [22]. As the result, only in the case of one parent the ES has the potential to realize nearly optimal mutation strengths.

Table 1. Important Symbols and Abbreviations

N	dimensionality of the search space
μ	number of parents
λ	number of offspring
$\mathcal{N}(\mu, \sigma)$	Normal distribution with mean μ and standard deviation σ
Φ	cumulative density function of standard normal distribution
ψ	self-adaptation response function (SAR)
φ_x	progress rate parallel to axis direction
φ_R	progress rate (distance to axis)
φ_R^*	progress rate (distance to axis) normalized w.r.t. N
φ_R^*	progress rate (distance to axis) normalized w.r.t. N
ΔQ^*	quality change normalized w.r.t. N
τ	learning rate; parameter of log-normal distribution
d	ridge parameter: weighting constant of embedded sphere
α	ridge parameter: degree of the ridge
R	distance to ridge axis
x	position on axis
q	abbreviation for $\alpha d R^{\alpha-1}$
ς^*	mutation strength normalized w.r.t. N
σ^*	abbreviation for $\langle \varsigma^{*(g)} \rangle$
$\varsigma_{\psi 0}^*$	zero of the SAR
$\varsigma_{\varphi R}^*$	zero of the progress rate φ_R
ς_{st}^*	stat. mutation strength
R_{st}	stat. distance to axis
$\varphi_{x st}^*$	stat. progress rate parallel to axis
$c_{\mu/\mu,\lambda}$	progress coefficient
$e_{\mu,\lambda}^{\alpha,\beta}$	generalized progress coefficient

4   Conclusion and Outlook

Self-adaptive intermediate ES show very different behaviors on the sharp and parabolic ridge. As we have seen, for $\alpha > 1$, the ES fluctuates at a stationary distance from the ridge with a positive mutation strength. As a result, there is progress in axis direction. However, the mutation strength is far from optimal for the overall fitness gain on the ridge. The ES is deceived by the sphere-model part of the ridge. Due to the influence of the distance on the SAR, the ES is not able to optimize the subgoals equally well. The SAR strives to maintain a positive mutation strength for decreasing distances which hinders a convergence towards the axis. As already pointed out in [14], the case of $\alpha > 1$ closely resembles the situation in the noisy sphere model where the ES is unable to converge to the optimizer and remains then on average at a certain distance to the optimizer.

In the case of the parabolic ridge, recombination appears to have no advantage compared to non-recombinative $(1, \lambda)$-ES. If μ is chosen to the usual rules, i.e, as long as $\mu \leq \lambda/2$, the mutation strength and the progress are lowered.

In the case of the sharp ridge, the ES either converges prematurely or enlarges the distance to the axis. This depends on the choice of the d-parameter with respect to the population parameters μ and λ. Additionally, it can be shown that even if self-adaptive ES do not converge prematurely, their travel speed is not optimal since self-adaptation realizes mutation strengths too small. In addition, increasing μ from 1 upwards lowers the critical d-factor for premature convergence.

This analysis must be extended to include the fluctuation parts of the evolution equations and the N-dependent versions of the progress rates and SAR. Additionally, the effects of noisy fitness evaluation should be investigated. And finally, a comparison with other adaptation schemes are of interest.

References

1. Rechenberg, I.: Evolutionsstrategie: Optimierung technischer Systeme nach Prinzipien der biologischen Evolution. Frommann-Holzboog Verlag, Stuttgart (1973)
2. Ostermeier, A., Gawelczyk, A., Hansen, N.: A derandomized approach to self-adaptation of evolution strategies. Evolutionary Computation 2(4), 369–380 (1995)
3. Hansen, N., Ostermeier, A.: Completely derandomized self-adaptation in evolution strategies. Evolutionary Computation 9(2), 159–195 (2001)
4. Schwefel, H.-P.: Adaptive Mechanismen in der biologischen Evolution und ihr Einfluß auf die Evolutionsgeschwindigkeit. Technical report, Technical University of Berlin. Abschlußbericht zum DFG-Vorhaben Re 215/2 (1974)
5. Bienvenüe, A., François, O.: Global convergence for evolution strategies in spherical problems: Some simple proofs and difficulties. Theoretical Computer Science 308, 269–289 (2003)
6. Auger, A.: Convergence results for the $(1, \lambda)$-SA-ES using the theory of ϕ-irreducible Markov chains. Theoretical Computer Science 334, 35–69 (2005)
7. Hart, W., DeLaurentis, J., Ferguson, L.: On the convergence of an implicitly self-adaptive evolutionary algorithm on one-dimensional unimodal problems. IEEE Transactions on Evolutionary Computation (2003) (to appear)

8. Hart, W.E.: Convergence of a discretized self-adaptive evolutionary algorithm on multi-dimensional problems (2003) (submitted)
9. Semenov, M.: Convergence velocity of evolutionary algorithms with self-adaptation. In: GECCO 2002, pp. 210–213 (2002)
10. Semenov, M., Terkel, D.: Analysis of convergence of an evolutionary algorithm with self-adaptation using a stochastic Lyapunov function. Evolutionary Computation 11(4), 363–379 (2003)
11. Beyer, H.-G.: Toward a theory of evolution strategies: Self-adaptation. Evolutionary Computation 3(3), 311–347 (1996)
12. Beyer, H.-G.: On the performance of $(1, \lambda)$-evolution strategies for the ridge function class. IEEE Transactions on Evolutionary Computation 5(3), 218–235 (2001)
13. Oyman, A.I., Beyer, H.-G., Schwefel, H.-P.: Analysis of a simple ES on the parabolic ridge. Evolutionary Computation 8(3), 249–265 (2000)
14. Beyer, H.-G.: Estimating the steady-state of CSA-ES on ridge functions. The Theory of Evolutionary Algorithms, Dagstuhl Seminar, Wadern, Germany (February 2004)
15. Arnold, D.V., Beyer, H.-G.: Evolution strategies with cumulative step length adaptation on the noisy parabolic ridge. Technical Report CS-2006-02, Dalhousie University, Faculty of Computer Science (2006)
16. Beyer, H.-G., Schwefel, H.-P.: Evolution strategies: A comprehensive introduction. Natural Computing 1(1), 3–52 (2002)
17. Bäck, T.: Self-adaptation. In: Bäck, T., Fogel, D., Michalewicz, Z. (eds.) Handbook of Evolutionary Computation, pp. C7.1:1–C7.1:15 Oxford University Press, New York (1997)
18. Arnold, B.C., Balakrishnan, N., Nagaraja, H.N.: A First Course in Order Statistics. Wiley, New York (1992)
19. Arnold, D.V.: Noisy Optimization with Evolution Strategies. Kluwer Academic Publishers, Dordrecht (2002)
20. Beyer, H.-G.: The Theory of Evolution Strategies. In: Natural Computing Series, Springer, Heidelberg (2001)
21. Herdy, M.: Reproductive isolation as strategy parameter in hierarchically organized evolution strategies. In: Männer, R., Manderick, B. (eds.) Parallel Problem Solving from Nature, vol. 2, pp. 207–217. Elsevier, Amsterdam (1992)
22. Meyer-Nieberg, S., Beyer, H.-G.: On the analysis of self-adaptive recombination strategies: First results. In: McKay, B., et al. (eds.) Proc. 2005 Congress on Evolutionary Computation (CEC'05), Edinburgh, UK, Piscataway, NJ, pp. 2341–2348. IEEE Press, NJ, New York (2005)
23. Braun, M.: Differential Equations and their Applications. Springer, Heidelberg (1998)
24. Beyer, H.-G., Arnold, D.V.: The steady state behavior of $(\mu/\mu_I, \lambda)$-ES on ellipsoidal fitness models disturbed by noise. In: Cantú-Paz, E., Foster, J.A., Deb, K., Davis, L., Roy, R., O'Reilly, U.-M., Beyer, H.-G., Kendall, G., Wilson, S.W., Harman, M., Wegener, J., Dasgupta, D., Potter, M.A., Schultz, A., Dowsland, K.A., Jonoska, N., Miller, J., Standish, R.K. (eds.) GECCO 2003. LNCS, vol. 2724, pp. 525–536. Springer, Heidelberg (2003)
25. Beyer, H.-G., Meyer-Nieberg, S.: Self-adaptation of evolution strategies under noisy fitness evaluations. Genetic Programming and Evolvable Machines (2006) (accepted)

A   The Density Function of an Offspring

Let us consider the quality change of an offspring l based on the centroid $\langle \mathbf{y} \rangle$ of the parent population

$$Q := F(\mathbf{y}^l) - F(\langle \mathbf{y} \rangle) = y_1^l - \langle y_1 \rangle - d(r^\alpha - R^\alpha)$$
$$=: z_x - d(r^\alpha - R^\alpha) \tag{31}$$

where $z_x := y_1^l - \langle y_1 \rangle$ denotes the change in the first component of the vector, whereas $R := \left(\sum_{k=2}^N (\langle y_k \rangle)^2 \right)^{1/2}$ denotes the centroid's distance to the ridge and ridge $r := \left(\sum_{k=2}^N (y_k^l)^2 \right)^{1/2}$ gives the distance of the offspring. In order to derive the cumulative density function (cdf) and probability density function (pdf) of an offspring several steps are needed:

1. Note, the ridge function is normalized $f(\mathbf{y}) = y_1 - d \left(\sum_{i=2}^N y_i^2 \right)^{\alpha/2} =: x - dR^\alpha$. Thus, $z_x = x - \langle x \rangle$ is the change of the first component of the object vector and obeys a $\mathcal{N}(0, \sigma)$-distribution.
2. The change $r - R$ is small. Under this assumption consider the Taylor series expansion of $f(r) = r^\alpha$ around R, $T_f(r) = R^\alpha + \alpha R^{\alpha-1}(r - R) + \mathcal{O}((r - R)^2)$. Provided that the contributions of the quadratic (and higher) terms can be neglected, the fitness change simplifies to $Q = x - \langle x \rangle - d\alpha R^{\alpha-1}(r - R) + \mathcal{O}((r - R)^2)$. Note the assumption above is only necessary to treat the case of general α. In the case of $\alpha = 1$, there is no quadratic term. In the case of $\alpha = 2$, it is possible to treat r^2 directly by the usual decomposition (see below). So, in the case of the sharp and the parabolic ridge the assumption is not required.
3. Consider the $(N-1)$-dimensional system $(y_2, \ldots, y_N)^{\mathrm{T}}$. An offspring is created by adding a mutation vector $\mathbf{z}$ to the parental vector $\mathbf{R}$, i.e. $\mathbf{r} = \mathbf{R} + \mathbf{z}$. Switching to a coordinate system with origin in $\mathbf{R}$, we can decompose $\mathbf{z}$ into two parts $-z_R \mathbf{e}_R + \mathbf{h}$ where $\mathbf{e}_R := \mathbf{R}/R$ and $\mathbf{h}$ is perpendicular to $\mathbf{R}$. This decomposition is similar to the decomposition in the case of the sphere model [20]. Therefore, the $\mathbf{r}$-vector can re-written as $\mathbf{r} = \mathbf{R} - z_R \mathbf{e}_R + \mathbf{h}$ and its length as $r = \|\mathbf{r}\| = \sqrt{(\mathbf{R} - z_R \mathbf{e}_R)^2 + h^2}$.
4. The distributions of the components of the $\mathbf{r}$-vector remains to be addressed. Due to the isotropy of the mutations used, the first component z_R will be assumed to be the second component of the object vector $\mathbf{y}$. It is therefore $\mathcal{N}(0, \sigma)$-distributed. The remaining sum $h^2 = \sum_{i=3}^N y_i^2$ consists of the squares of $N - 2$ normally distributed random variables and is χ^2-distributed. A χ^2-distribution may be modeled using a normal distribution. Considering sufficiently large N, h^2 is $\mathcal{N}(N\sigma^2, \sqrt{2N}\sigma^2)$ distributed.
5. Consider the square root

$$f(z_R, h_R) = \sqrt{(R - z_R)^2 + h^2} = \sqrt{R^2 - 2Rz_R + z_R^2 + h^2} \tag{32}$$

which can be rewritten as

$$f(z_R, h_R) = R\sqrt{1 - \frac{2}{R}z_R + \frac{z_R^2}{R^2} + \frac{h^2}{R^2}}$$

$$= R\sqrt{1 - 2\left(\frac{z_R}{R} - \frac{z_R^2}{2R^2} - \frac{h^2}{2R^2}\right)}. \tag{33}$$

Provided that $z_R \ll R$, $h \ll R$ hold, the root can be expanded into a Taylor series around zero and cut off after the very first term giving $f(z_R, h_R) = R(1 - z_R/R + z_R^2/(2R^2) + h^2/(2R^2))$. Provided that $z_R^2/(2R^2) \ll 1$, the term may be neglected.

6. Let us treat the case of $\alpha = 2$ separately. Here, we have $r^2 = (R - z_R)^2 + h^2 = R^2 + 2R^2(-z_R/R + z_R^2/(2R^2) + h^2/(2R^2))$ and only require that the contribution of the square of z_R is negligible.

As already pointed out in [14] the resulting quality change

$$Q = z_x - d\alpha R^{\alpha-1}\left(R(1 - \frac{z_R}{R} + \frac{h^2}{2R^2}) - R\right)$$
$$= z_x + d\alpha R^{\alpha-1}\left(z_R - \frac{h^2}{2R}\right) \tag{34}$$

is very similar to that of a noisy sphere with z_x in the role of the noise term.

The cumulative density function (cdf) and the probability density function (pdf) of an offspring can now be easily given as

$$P(Q) = \Phi\left(\frac{Q + q\frac{N}{2R}\sigma^2}{\sqrt{\sigma^2(1+q^2) + q^2\frac{N}{2R^4}\sigma^4}}\right) \tag{35}$$

and

$$p(Q) = \frac{\exp\left(-\frac{1}{2}\left(\frac{Q+q\frac{N}{2R}\sigma^2}{\sqrt{\sigma^2(1+q^2)+q^2\frac{N}{2R^4}\sigma^2}}\right)^2\right)}{\sqrt{2\pi}\sqrt{\sigma^2(1+q^2) + q^2\frac{N}{2R^4}\sigma^4}} \tag{36}$$

with $q = d\alpha R^{\alpha-1}$ (7). Introducing the normalizations $Q^* = Q/N$ and $\sigma^* = \sigma/N$, the pdf and cdf change to

$$P(Q^*) = \Phi\left(N\frac{\frac{Q^*}{N} + q\frac{\sigma^{*2}}{2RN}}{\sqrt{\sigma^{*2}(1+q^2) + q^2\frac{\sigma^{*4}}{2N^2R^4}}}\right)$$
$$= \Phi\left(\frac{Q^* + q\frac{\sigma^{*2}}{2R}}{\sqrt{\sigma^{*2}(1+q^2) + q^2\frac{\sigma^{*4}}{2N^2R^4}}}\right) \tag{37}$$

and

$$p(Q^*) = N\frac{\exp\left(-\frac{1}{2}\left(\frac{Q^*+q\frac{\sigma^{*2}}{2R}}{\sqrt{\sigma^{*2}(1+q^2)+q^2\frac{\sigma^{*4}}{2N^2R^4}}}\right)^2\right)}{\sqrt{2\pi}\sqrt{\sigma^{*2}(1+q^2) + q^2\frac{\sigma^{*2}}{2N^2R^4}}} \tag{38}$$

The expression $p(Q)\,dQ$ is equal to $1/N\,p(Q^*)\,dQ^* = p^*(Q^*)\,dQ^*$. In the case of $N \to \infty$, the components stemming from the distance's perpendicular part vanish leading to

$$P(Q^*) = \Phi\left(\frac{Q^* + q\frac{\sigma^{*2}}{2R}}{\sqrt{\sigma^{*2}(1+q^2)}}\right) \tag{39}$$

and

$$p^*(Q^*) = \frac{\exp\left(-\frac{1}{2}\left(\frac{Q^*+q\frac{\sigma^{*2}}{2R}}{\sqrt{\sigma^{*2}(1+q^2)}}\right)^2\right)}{\sqrt{2\pi}\sqrt{\sigma^{*2}(1+q^2)}} \tag{40}$$

which will be used in the remainder of this paper.

B   The Progress Rates

The following lemma is taken directly from [15, p.6]:

Lemma 1. *Let $Y_1, Y_2, \ldots, Y_\lambda$ be λ independent standard normally distributed random variables and let $Z_1, Z_2, \ldots, Z_\lambda$ be λ independent normally distributed random variables with zero mean and variance θ^2. Then, defining $X_l = Y_l + Z_l$ for $l = 1, \ldots, \lambda$ and ordering the sample members by nondecreasing values of the X variates, the expected value of the arithmetic mean of those μ of the Y_l with the largest associated values of X_l is*

$$\langle Y \rangle = \frac{c_{\mu/\mu,\lambda}}{\sqrt{1 + \theta^2}}. \tag{41}$$

Lemma 1 can be used to determine the progress rates. Note, the same decomposition as in Appendix A applies: The quality change of an offspring is given by

$$Q = z_x + q z_R - \frac{q}{2R} h^2. \tag{42}$$

The random variables z_x and z_R are normally distributed with mean zero and standard deviation σ. Similarly, the random variable h^2 may be assumed to be normally distributed with mean $N\sigma^2$ and standard deviation $\sqrt{2N}\sigma^2$ if N is sufficiently large. In the following, we will switch to standard normally distributed random variables u_*:

$$Q = \sigma u_x + q\sigma u_R - \frac{q}{2R}\sqrt{2N}\sigma^2 u_{h^2} - \frac{q}{2R}N\sigma^2. \tag{43}$$

Let us start with the axial progress

$$\varphi_x = \mathrm{E}\left[\langle x^{(g+1)} \rangle - \langle x^{(g)} \rangle\right] = \mathrm{E}[\langle z_x \rangle] = \sigma \mathrm{E}[\langle u_x \rangle]. \tag{44}$$

The expectation can be determined using Lemma 1. Note, the addend $q/(2R)N\sigma^2$ in (43) does not influence the selection since it is the same for all offspring. The corresponding normally distributed variables Z_l of Lemma 1 are defined by

$$Z_l = \frac{q}{\sigma}\sigma u_R - \frac{q}{2\sigma R}\sqrt{2N}\sigma^2 u_{h^2} = q\sqrt{1 + \frac{N}{2R^2}\sigma^2}\, \mathcal{N}_l(0,1). \tag{45}$$

where $\mathcal{N}_l(0,1)$ denotes a standard normally distributed random variable. Note, the sum of two normally distributed random variables is again a normally distributed random variable. Therefore, Lemma 1 gives

$$\varphi_x = \frac{c_{\mu/\mu,\lambda}\sigma^2}{\sqrt{\sigma^2(1 + q^2) + q^2 \frac{N}{2R^2}\sigma^4}}. \tag{46}$$

The progress (not normalized and normalized) towards the axis is defined as

$$\varphi_R := \mathrm{E}[R - r] = R\mathrm{E}\left[1 - \sqrt{\left(1 - \frac{\langle z_R \rangle}{R}\right)^2 + \frac{\langle h \rangle^2}{R^2}}\right]$$

$$\varphi_R^* := N\mathrm{E}[R - r] = RN\mathrm{E}\left[1 - \sqrt{\left(1 - \frac{\langle z_R \rangle}{R}\right)^2 + \frac{\langle h \rangle^2}{R^2}}\right]. \tag{47}$$

To continue, we use the results obtained in [20] and [15]:

1. It was shown in [20] that

$$\varphi_R^* = NR(1 - \sqrt{\left(1 - \frac{\overline{\langle z_R \rangle}}{R}\right)^2 + \frac{\overline{\langle h \rangle^2}}{R^2}}) + \mathcal{O}\left(\frac{1}{\sqrt{N}}\right). \tag{48}$$

2. To determine the expectation of the central component Lemma 1 can be used. The determination is completely analogous to the determination of $E[\langle z_x \rangle]$. Only the roles of z_R and z_x are reversed:

$$\overline{\langle z_R \rangle} = \frac{c_{\mu/\mu,\lambda} q \sigma^2}{\sqrt{\sigma^2(1 + q^2) + q^2 \frac{N}{2R^2} \sigma^4}}. \tag{49}$$

3. In the case of the lateral component, the expectation over the square of the sum of μ vectors must be taken. Since the random vectors $\mathbf{h}_{m;\lambda}$ are independent [20], $E[\mathbf{h}_{m;\lambda}^{\mathrm{T}} \mathbf{h}_{l;\lambda}] = 0$ holds for $m \neq l$. The expectation

$$\overline{\langle \mathbf{h} \rangle^2} = \frac{\overline{\langle \mathbf{h}^2 \rangle}}{\mu} \tag{50}$$

remains. Remember, the random variable h^2 of each offspring is also a normally distributed random variable with mean $N\sigma^2$ and standard deviation $\sqrt{2N}\sigma^2$

$$\frac{\overline{\langle \mathbf{h}^2 \rangle}}{\mu} = \frac{\sqrt{2N}}{\mu} \sigma^2 \overline{\langle u_{h^2} \rangle} + \frac{N}{\mu} \sigma^2. \tag{51}$$

Let us now consider $\overline{\langle u_{h^2} \rangle}$. Using (43), the corresponding Z_l of Lemma 1 read

$$Z_l = \frac{\sigma}{\frac{q}{2R}\sqrt{2N\sigma^2}} u_x + \frac{q\sigma}{\frac{q}{2R}\sqrt{2N\sigma^2}} u_z = \frac{\sqrt{1+q^2}\sigma}{\sqrt{2N\sigma^2}\frac{q}{2R}} \mathcal{N}_l(0, 1). \tag{52}$$

Taking note of the sign in (43), this leads to

$$\overline{\langle u_{h^2} \rangle} = -\frac{\frac{q}{2R}\sqrt{2N}c_{\mu/\mu,\lambda}\sigma^2}{\sqrt{\sigma^2(1 + q^2) + q^2 \frac{N}{2R^2}\sigma^4}}. \tag{53}$$

By plugging (53) into (51),

$$\frac{\overline{\langle \mathbf{h}^2 \rangle}}{\mu} = -\frac{c_{\mu/\mu,\lambda} q \frac{N}{R} \sigma^4}{\mu \sqrt{\sigma^2(1 + q^2) + q^2 \frac{N}{2R^2}\sigma^4}} + \frac{N}{\mu}\sigma^2 \tag{54}$$

is obtained. Introducing again the normalization $\sigma^* = N\sigma$, (54) changes to

$$\frac{\overline{\langle \mathbf{h}^2 \rangle}}{\mu} = -\frac{c_{\mu/\mu,\lambda} \frac{q}{RN^2} \sigma^{*4}}{\mu \sqrt{\sigma^{*2}(1 + q^2) + q^2 \frac{\sigma^{*4}}{2R^2 N}}} + \frac{\sigma^{*2}}{\mu N} \tag{55}$$

and (49) becomes

$$\overline{\langle z_R \rangle} = \frac{c_{\mu/\mu,\lambda} q \frac{\sigma^{*2}}{N}}{\sqrt{\sigma^{*2}(1+q^2) + \frac{q^2}{2NR^2}\sigma^{*4}}}. \tag{56}$$

The results (55) and (56) are then inserted into the lateral progress rate (48).

4. Using Taylor series expansions (see [20, p.215]) for (48)) and the resulting expressions, it can be shown that for $N \to \infty$

$$\varphi_R^* = \frac{q\sigma^*}{\sqrt{1+q^2}} c_{\mu/\mu,\lambda} - \frac{\sigma^{*2}}{2R\mu} \tag{57}$$

is obtained. Although the calculations are straightforward, they are lengthy and of purely technical nature. Therefore, they are omitted at this point. Equation (57) will serve as an approximate formula for finite dimensional search spaces. Note, the progress rates (46) and (57) were obtained for the case $\tau = 0$ and thus only applicable for small values of τ.

C   The Self-adaptation Response

C.1   The First Order Self-adaptation Response

To simplify the notations, we will again set $\sigma^* := \langle \varsigma^{*(g)} \rangle$, $R := r^{(g)}$, and $X := x^{(g)}$. The self-adaptation response denotes the relative expected change of the mean of the population's mutation strength, i.e., $\psi = \mathrm{E}[(\langle \varsigma^* \rangle - \sigma^*)/\sigma^*]$ or

$$\psi(\sigma^*, x^{(g)}, r^{(g)}) = \frac{1}{\mu} \sum_{m=1}^{\mu} \mathrm{E}\left[\frac{\varsigma_{m;\lambda}^* - \sigma^*}{\sigma^*}\right]$$

$$= \int_0^\infty \left(\frac{\varsigma^* - \sigma^*}{\sigma^*}\right) p_{m;\lambda}(\varsigma^*, \sigma^*, x^{(g)}, r^{(g)}) \, \mathrm{d}\varsigma^*. \tag{58}$$

The mutation strength $\varsigma_{m;\lambda}^*$ is the mutation strength that is connected with the offspring with the mth highest quality change in λ trials. To continue, we need its density function (pdf)

$$p_{m;\lambda}(\varsigma^*, \sigma^*, X, R) = p_{\sigma^*}(\varsigma^*|\sigma^*) \frac{\lambda}{(m-1)!(\lambda-m)!} \int_{-\infty}^\infty p(Q^*|\varsigma^*, X, R)$$

$$\times P(Q^*|\sigma^*, X, R)^{\lambda-m}\left(1 - P(Q^*|\sigma^*, X, R)\right)^{m-1} \mathrm{d}Q^*. \tag{59}$$

Note, it is generally not possible to obtain analytical closed solutions of (58). Therefore, we will resort to approximate expressions. The first concerns the cumulative density function (cdf) $P(Q^*|\sigma^*, X, R)$ which is given as the expectation

$$P(Q^*|\sigma^*, X, R) = \int_0^\infty P(Q^*|\varsigma^*, X, R) p_{\sigma^*}(\varsigma^*|\sigma^*) \, \mathrm{d}\varsigma^*$$

$$= \int_0^\infty \Phi\left(\frac{Q^* + q\frac{\varsigma^{*2}}{2R}}{\sqrt{\varsigma^{*2}(1+q^2)}}\right) p_{\sigma^*}(\varsigma^*|\sigma^*) \, \mathrm{d}\varsigma^* \tag{60}$$

over the ς^*-range. The integral expression can be simplified if some assumptions are met [20]. In the case of a log-normal operator, it can be shown that

$$P(Q^*|\sigma^*, X, R) = \Phi\left(\frac{Q^* + q\frac{\sigma^{*2}}{2R}}{\sqrt{\sigma^{*2}(1 + q^2)}}\right) + \mathcal{O}(\tau^2). \tag{61}$$

Assuming that $\tau \ll 1$ holds, the τ-dependent part can be neglected. In the following, we will aim at a simplification of the formulae. The first step is a substitution of the argument in (61) with

$$z = -\frac{Q^* + q\frac{\sigma^{*2}}{2R}}{\sqrt{\sigma^{*2}(1 + q^2)}} \Rightarrow Q^* = -\sqrt{\sigma^{*2}(1 + q^2)}z - q\frac{\sigma^{*2}}{2R}. \tag{62}$$

The density function $p(Q^*|\varsigma^*, X, R)$ changes accordingly to

$$p(Q^*|\varsigma^*, X, R)\, dQ^* = \frac{1}{\sqrt{2\pi}} \frac{e^{-\frac{1}{2}\left(\frac{Q^* + q\frac{\varsigma^{*2}}{2R}}{\sqrt{\varsigma^{*2}(1 + q^2)}}\right)^2}}{\sqrt{\varsigma^{*2}(1 + q^2)}}\, dQ^*$$

$$\Rightarrow -p(z|\varsigma^*, X, R)\, dz = -\frac{1}{\sqrt{2\pi}} \frac{\sigma^*}{\varsigma^*}$$

$$\times \exp\left(-\frac{1}{2}\left(\frac{\sigma^*}{\varsigma^*}z - \frac{\frac{q}{2R}(\varsigma^{*2} - \sigma^{*2})}{\sqrt{\varsigma^{*2}(1 + q^2)}}\right)^2\right)\, dz$$

$$=: -\frac{1}{\sqrt{2\pi}}G(\varsigma^*)F(\varsigma^*, z)\, dz \tag{63}$$

where $F(\varsigma^*, z)$ denotes the exponential function and $G(\varsigma^*) = \sigma^*/\varsigma^*$. Let us come back to (58) which can now be given as

$$\psi(\sigma^*, X, R) = \int_0^\infty \left(\frac{\varsigma^* - \sigma^*}{\sigma^*}\right) p_{\sigma^*}(\varsigma^*|\sigma^*)$$

$$\times \int_{-z_0}^{z_e} p(z|\varsigma^*, X, R)\frac{1}{\mu}\sum_{m=1}^\mu \frac{\lambda}{(m-1)!(\lambda - m)!}$$

$$\times \Phi(z)^{\lambda - m}\left(1 - \Phi(z)\right)^{m-1}\, dz\, d\varsigma^*. \tag{64}$$

Let us first consider $1/\mu \sum_{m=1}^\mu \lambda/[(m - 1)!(\lambda - m)!](1 - \Phi(z))^{\lambda - m}\Phi(z)^{m-1}$ which represents a regularized incomplete beta function [20]. Therefore, it can be substituted by the integral $1/[(\lambda - \mu - 1)!(\mu - 1)!] \int_0^{1 - \Phi(z)} x^{\lambda - \mu - 1}(1 - x)^{\mu - 1}\, dx$. Equation (64) changes to

$$\psi(\sigma^*, X, R) = \frac{\lambda!}{\mu(\lambda - \mu - 1)!(\mu - 1)!} \int_0^\infty \left(\frac{\varsigma^* - \sigma^*}{\sigma^*}\right) p_{\sigma^*}(\varsigma^*|\sigma^*)$$

$$\times \int_{-z_0}^{z_e} p(z|\varsigma^*, X, R)$$

$$\times \int_0^{1 - \Phi(z)} x^{\lambda - \mu - 1}(1 - x)^{\mu - 1}\, dx\, dz\, d\varsigma^*. \tag{65}$$

In the next step, we change the integration order of the integration over z and x

$$\psi(\sigma^*, X, R) = \frac{\lambda!}{\mu(\lambda - \mu - 1)!(\mu - 1)!} \int_0^\infty \left(\frac{\varsigma^* - \sigma^*}{\sigma^*}\right) p_{\sigma^*}(\varsigma^*|\sigma^*)$$

$$\times \int_0^1 \int_0^{\Phi^{-1}(1-x)} p(z|\varsigma^*, X, R) \, dz$$

$$\times x^{\lambda - \mu - 1}(1 - x)^{\mu - 1} \, dx \, d\varsigma^*. \tag{66}$$

Setting $x = (1 - \Phi(t))$, (66) changes to

$$\psi(\sigma^*, X, R) = \frac{\lambda!}{\mu(\lambda - \mu - 1)!(\mu - 1)!} \int_0^\infty \left(\frac{\varsigma^* - \sigma^*}{\sigma^*}\right) p_{\sigma^*}(\varsigma^*|\sigma^*)$$

$$\times \int_0^\infty \int_0^t p(z|\varsigma^*, X, R) \, dz$$

$$\times \left(1 - \Phi(t)\right)^{\lambda - \mu - 1} \Phi(t)^{\mu - 1} e^{-\frac{t^2}{2}} \, dt \, d\varsigma^*. \tag{67}$$

In order to continue, the functions G and F introduced in (63) will be expanded into their Taylor series' T_G and T_F around σ^*. From this point we will consider

$$p(z|\varsigma^*, X, R) \, dz = T_G(\varsigma^*) T_F(\varsigma^*) \, dz = \sum_{k=0}^\infty \frac{\partial^k}{\partial \varsigma^{*k}} G(\varsigma^*)\big|_{\varsigma^* = \sigma^*} \frac{(\varsigma^* - \sigma^*)^k}{k!}$$

$$\times \sum_{l=0}^\infty \frac{\partial^l}{\partial \varsigma^{*l}} F(\varsigma^*, z)\big|_{\varsigma^* = \sigma^*} \frac{(\varsigma^* - \sigma^*)^l}{l!} \, dz. \tag{68}$$

To this end, the functions' derivatives must be derived (see Appendix C.2). The question remains when the Taylor series may be cut off without introducing a serious approximation error. The approximability depends on the assumption of $\tau \ll 1$. The consequences are the following: The expectation of $(\varsigma^* - \sigma^*)^k$ leads to $\sigma^{*k}(-1)^k \sum_{l=0}^k \binom{k}{l}$ $(-1)^l \exp(l^2\tau^2/2)$. The exponential function can be expressed as $\exp(l^2\tau^2/2) = \sum_{m=0}^\infty l^{2m}\tau^{2m}/m!$. Therefore, the expectation of $(\varsigma^* - \sigma^*)^k$ only contains even powers of τ. Because of $\tau \ll 1$ it is possible to neglect higher order terms of τ, (i.e., τ^{2m} with $m \geq 2$) without introducing serious approximation errors.

The question remains whether and when we can cut off the Taylor series in $(\varsigma^* - \sigma^*)$. It can be shown that $E[(\varsigma^* - \sigma^*)^k]$ only contains τ^{2m} terms which fulfill $k < 2m + 1$. In other words, terms with τ^2 only appear for the first and second moment. Once we have $k \geq 3$, the lowest power of τ appearing is four (see e.g. [25]).

This has two effects: First, we can cut off the series after the quadratic term and introduce an error term of order $\mathcal{O}(\tau^4)$. Second, since we multiply the series' with $(\varsigma^* - \sigma^*)/\sigma^*$ all terms of the result with $(\varsigma^* - \sigma^*)^3$ and $(\varsigma^* - \sigma^*)^4$ can also be neglected. That is, all contributions from the quadratic or higher terms of the original series enter the error term and only the contributions of the first derivatives need to be taken into account.

We will first address the integration over z. It can be shown that the Taylor series of F is a polynomial in z. Let f_k denote all components of the Taylor series of F which are connected with z^k. Equation (63) is of the form

$$p(z|\varsigma^*, X, R)\,dz = \frac{1}{\sqrt{2\pi}}G(\varsigma^*)F(\varsigma^*, z)\,dz = \frac{1}{\sqrt{2\pi}}G(\varsigma^*)T_F(\varsigma^*, z)\,dz$$

$$= \frac{e^{-\frac{z^2}{2}}}{\sqrt{2\pi}}G(\varsigma^*)\Big(f_0 + f_1 z + f_2 z^2 + f_3 z^3 + f_4 z^4$$

$$+\mathcal{R}(z, \varsigma^*)\Big)\,dz. \tag{69}$$

The integration over z is straightforward (see [20, p.331f])

$$\int_0^t p(z|\varsigma^*, X, R)\,dz = G(\varsigma^*)\int_0^t \frac{e^{-\frac{z^2}{2}}}{\sqrt{2\pi}}\Big(f_0 + f_1 z + f_2 z^2 + f_3 z^3 + f_4 z^4\Big)\,dz.$$

$$= \frac{G(\varsigma^*)}{\sqrt{2\pi}}\Big(\sqrt{2\pi}f_0\Phi(t) - f_1 e^{-\frac{z^2}{2}}$$

$$+f_2\Big(\sqrt{2\pi}\Phi(t) - te^{-\frac{t^2}{2}}\Big) - f_3 e^{-\frac{t^2}{2}}\Big(2 + t^2\Big)$$

$$+3\sqrt{2\pi}f_4\Phi(t) - f_4 e^{-\frac{t^2}{2}}\Big(3t + t^3\Big)\Big)$$

$$= G(\varsigma^*)\Big((f_0 + f_2 + 3f_4)\Phi(t) - \frac{e^{-\frac{z^2}{2}}}{\sqrt{2\pi}}$$

$$\times\Big(f_1 + 2f_3 + (f_2 + 3f_4)t + f_3 t^2 + f_4 t^3\Big)\Big). \tag{70}$$

The next step is the integration over t

$$I_t(\varsigma^*) = \frac{G(\varsigma^*)\lambda!}{\mu(\lambda - \mu - 1)!(\mu - 1)!}\int_0^\infty \Big((f_0 + f_2 + 3f_4)\Phi(t) - \frac{e^{-\frac{z^2}{2}}}{\sqrt{2\pi}}$$

$$\times\Big(f_1 + 2f_3 + (f_2 + 3f_4)t + f_3 t^2 + f_4 t^3\Big)\Big)$$

$$\times\Big(1 - \Phi(t)\Big)^{\lambda-\mu-1}\Phi(t)^{\mu-1}e^{-\frac{t^2}{2}}\,dt\,d\varsigma^* \tag{71}$$

which cannot be solved analytically. Instead, we use the so-called generalized progress coefficients (8). Equation (71) changes to

$$I_t(\varsigma^*) = G(\varsigma^*)\Big(f_0 + f_2 + 3f_4 - \Big((f_1 + 2f_3)e_{\mu,\lambda}^{1,0}$$

$$-(f_2 + 3f_4)e_{\mu,\lambda}^{1,1} + f_3 e_{\mu,\lambda}^{1,2} - f_4 e_{\mu,\lambda}^{1,3}\Big)\Big)$$

$$= G(\varsigma^*)\Big(f_0 + f_2 + 3f_4 - (f_1 + 2f_3)c_{\mu/\mu,\lambda}$$

$$+(f_2 + 3f_4)e_{\mu,\lambda}^{1,1} - f_3 e_{\mu,\lambda}^{1,2} + f_4 e_{\mu,\lambda}^{1,3}\Big). \tag{72}$$

The integration over ς^* still remains. The coefficients f_k in (72) are polynomials in $(\varsigma^* - \sigma^*)$. As mentioned before, all contributions of terms higher than linear order can be neglected. Since it can be shown by calculating the derivatives of F that f_3 and f_4 contain only quadratic (or higher) terms, only f_0, f_1, and f_2 need to be taken into account

$$
I_t(\varsigma^*) = G(\varsigma^*)\Big(1 + f_{21}(\varsigma^* - \sigma^*) - f_{11}(\varsigma^* - \sigma^*)c_{\mu/\mu,\lambda}
$$

$$
+ f_{21}(\varsigma^* - \sigma^*)e_{\mu,\lambda}^{1,1} + \mathcal{O}\big((\varsigma^* - \sigma^*)^2\big)\Big)
$$

$$
= G(\varsigma^*)\Big(1 + (f_{21} - f_{11}c_{\mu/\mu,\lambda} + f_{21}e_{\mu,\lambda}^{1,1})(\varsigma^* - \sigma^*)
$$

$$
+ \mathcal{O}\big((\varsigma^* - \sigma^*)^2\big)\Big). \tag{73}
$$

with f_{kl} denoting the component of lth derivative of F associated with z^k. As mentioned before, we will expand the function G into its Taylor series $T_G(\varsigma^*)$ around σ^*, i.e., $T_G(\varsigma^*) = \sum_{k=0}^{\infty} G_k/(k!)(\varsigma^* - \sigma^*)^k = g_0 + g_1(\varsigma^* - \sigma^*) + \mathcal{O}\big((\varsigma^* - \sigma^*)^2\big)$ which gives

$$
I_t(\varsigma^*) = g_0 + (f_{21} - f_{11}c_{\mu/\mu,\lambda} + f_{21}e_{\mu,\lambda}^{1,1})g_0(\varsigma^* - \sigma^*)
$$

$$
+ g_1(\varsigma^* - \sigma^*) + \mathcal{O}\big((\varsigma^* - \sigma^*)^2\big). \tag{74}
$$

Inserting (74) into the SAR and taking the integration over ς^* gives

$$
\psi(\varsigma^*) = \int_0^{\infty} \left(\frac{\varsigma^* - \sigma^*}{\sigma^*}\right) I_t(\varsigma^*) p_{\sigma^*}(\varsigma^*, \sigma^*)\, d\sigma^*
$$

$$
= \int_0^{\infty} \left(g_0\left(\frac{\varsigma^* - \sigma^*}{\sigma^*}\right) + (f_{21} - f_{11}c_{\mu/\mu,\lambda} + f_{21}e_{\mu,\lambda}^{1,1})g_0\sigma^*\left(\frac{\varsigma^* - \sigma^*}{\sigma^*}\right)^2 \right.
$$

$$
\left. + g_1\sigma^{*2}\left(\frac{\varsigma^* - \sigma^*}{\sigma^*}\right)^2 + \mathcal{O}\big((\varsigma^* - \sigma^*)^3\big) \right) p_{\sigma^*}(\varsigma^*|\sigma^*)\, d\varsigma^*. \tag{75}
$$

The expectation of $(\varsigma^* - \sigma^*)^k$ can be easily computed (see e.g. [20] or [25]) As mentioned before all terms with $(\varsigma^* - \sigma^*)^k$ with $k > 3$ contain only expressions of $\mathcal{O}(\tau^4)$. We obtain

$$
\psi(\sigma^*) = g_0\frac{\tau^2}{2} + (f_{21} - f_{11}c_{\mu/\mu,\lambda} + f_{21}e_{\mu,\lambda}^{1,1}g_0 + g_1)\sigma^*\tau^2 + \mathcal{O}(\tau^4)
$$

$$
= \tau^2\Big(\frac{g_0}{2} + (f_{21} + g_1 - f_{11}c_{\mu/\mu,\lambda} + f_{21}e_{\mu,\lambda}^{1,1}g_0)\sigma^*\Big) + \mathcal{O}(\tau^4). \tag{76}
$$

The coefficients g_k and f_{kl} are obtained in Appendix C.2, Eqs. (87) and (88). Inserting (87) and (88), i.e. $g_0 = G(\sigma^*) = 1$, $g_1 = \partial/(\partial\varsigma^*)G(\varsigma^*)|_{\varsigma^*=\sigma^*} = -1/\sigma^*$, $f_{11} = -q/(R\sqrt{1+q^2})$, and $f_{21} = 1/\sigma^*$ into (76),

$$
\psi_{\infty}(\sigma^*) = \tau^2\left(\frac{1}{2} + e_{\mu,\lambda}^{1,1} - c_{\mu/\mu,\lambda}\sigma^*\sqrt{\frac{q^2}{R^2(1+q^2)}}\right) + \mathcal{O}(\tau^4) \tag{77}
$$

is obtained for $N \to \infty$ which will serve as an approximate equation for finite dimensional search spaces. Equation (77) was obtained under the assumption $\tau \ll 1$ which allows to neglect error terms of order τ^4 in the approximation.

C.2 Deriving the Derivatives

Concerning the variable ς^* the transformed density function (63)

$$p(z|\varsigma^*, X, R)\, dz = \frac{dz}{\sqrt{2\pi}} \frac{\sigma^*}{\varsigma^*}$$

$$\times \exp\left(-\frac{1}{2}\left(\frac{\sigma^*}{\varsigma^*} z - \frac{\frac{q}{2R}\left(\varsigma^{*2} - \sigma^{*2}\right)}{\varsigma^*\sqrt{1+q^2}} \right)^2 \right)$$

can be given as the general equation

$$p(z|\varsigma^*, X, R)\, dz = G(\varsigma^*)F(\varsigma^*)\, dz = \frac{g(\sigma^*)}{g(\varsigma^*)} e^{-\frac{f(\varsigma^*)^2}{2}}\, dz$$

$$= \frac{g(\sigma^*)}{g(\varsigma^*)} e^{-\frac{1}{2g(\varsigma^*)^2}\left(g(\sigma^*)z + K(h(\varsigma^*) - h(\sigma^*)) \right)^2}\, dz \qquad (78)$$

(see also Appendix C.1). In Appendix C.1, Eq.(68), the functions G and F are be expanded into their Taylor series' T_G and T_F around σ^*. To this end, the functions' derivatives are needed. The remainder of this section is devoted to this task. Let us start with the function

$$G(\varsigma^*) = \frac{g(\sigma^*)}{g(\varsigma^*)} \qquad (79)$$

where $g(x) = x$. The first derivative is easily given as

$$\frac{\partial}{\partial \varsigma^*} G(\varsigma^*) = -\frac{g(\sigma^*)g'(\varsigma^*)}{g(\varsigma^*)^2}. \qquad (80)$$

The function F is more complicated and depends additionally on the variable z. In its most general form F and its first derivative reads

$$F(\varsigma^*) = e^{-\frac{f(\varsigma^*)^2}{2}}$$

$$\frac{\partial}{\partial \varsigma^*} F(\varsigma^*) = -f'(\varsigma^*)f(\varsigma^*)e^{-\frac{f(\varsigma^*)^2}{2}} \qquad (81)$$

where the function f is composed as follows

$$f(\varsigma^*) = \frac{1}{g(\varsigma^*)}\left(g(\sigma^*)z + K(h(\varsigma^*) - h(\sigma^*)) \right). \qquad (82)$$

Its first derivatives in turn is

$$f'(\varsigma^*) = -\frac{g'(\varsigma^*)}{g(\varsigma^*)^2}\left(g(\sigma^*)z + K(h(\varsigma^*) - h(\sigma^*)) \right) + K\frac{h'(\varsigma^*)}{g(\varsigma^*)} \qquad (83)$$

As mentioned, we expand the functions F and G around the mean of the previous parental mutation strengths σ^*. Thus, we only need the derivatives at this point. This simplifies the equations significantly. In the case of the function f,

$$
f(\sigma^*) = \frac{1}{g(\sigma^*)}\Big(g(\sigma^*)z + K(h(\sigma^*) - h(\sigma^*))\Big) = z
$$

$$
f'(\sigma^*) = -\frac{g'(\sigma^*)}{g(\sigma^*)^2}\Big(g(\sigma^*)z + K(h(\sigma^*) - h(\sigma^*))\Big) + K\frac{h'(\sigma^*)}{g(\sigma^*)}
$$

$$
= -\frac{g'(\sigma^*)}{g(\sigma^*)}z + K\frac{h'(\sigma^*)}{g(\sigma^*)} \tag{84}
$$

are obtained. The function G and its first two derivatives become

$$
G(\sigma^*) = 1
$$

$$
\frac{\partial}{\partial\varsigma^*}G(\sigma^*) = -\frac{g'(\sigma^*)}{g(\sigma^*)}
$$

$$
\frac{\partial^2}{\partial\varsigma^{*2}}G(\sigma^*) = \frac{1}{g(\sigma^*)}\left(2\frac{g'(\sigma^*)^2}{g(\sigma^*)} - g''(\sigma^*)\right). \tag{85}
$$

The function F in turn becomes

$$
F(\sigma^*) = e^{-\frac{z^2}{2}}
$$

$$
\frac{\partial}{\partial\varsigma^*}F(\sigma^*) = -f'(\sigma^*)ze^{-\frac{z^2}{2}} = \left(\frac{g'(\sigma^*)}{g(\sigma^*)}z^2 + K\frac{h'(\sigma^*)}{g(\sigma^*)}z\right)e^{-\frac{z^2}{2}} \tag{86}
$$

We now need the derivatives of the functions g and h. Since $g(\varsigma^*) = \varsigma^*$, its first derivative is $g'(\varsigma^*) = 1$. The function h is given by $h(\varsigma^*) = \varsigma^{*2}$ which leads to $h'(\varsigma^*) = 2\varsigma^*$. The constant K (with respect to ς^*) reads $K = -q/(2R)$. Thus, we obtain

$$
F(\sigma^*) = e^{-\frac{z^2}{2}} \text{ and } \frac{\partial}{\partial\varsigma^*}F(\sigma^*) = \left(\frac{z^2}{\sigma^*} - \frac{q}{R\sqrt{1+q^2}}z\right)e^{-\frac{z^2}{2}} \tag{87}
$$

i.e., $f_{11} = -R\sqrt{q^2/(1+q^2)}$ and $f_{21} = 1/\sigma^*$ and

$$
G(\sigma^*) = 1 \text{ and } \frac{\partial}{\partial\varsigma^*}G(\sigma^*) = -\frac{1}{\sigma^*} \tag{88}
$$

which are used to determine the SAR's (76) coefficients in Appendix C.1.

Genericity of the Fixed Point Set for the Infinite Population Genetic Algorithm

Tomáš Gedeon[1], Christina Hayes[2], and Richard Swanson[1]

[1] Department of Mathematical Sciences, Montana State University
Bozeman, MT 59715
gedeon@math.montana.edu, swanson@math.montana.edu
[2] Department of Mathematics, Gettysburg Collge
Gettysburg, PA 17325
chayes@gettysburg.edu

Abstract. The infinite population model for the genetic algorithm, where the iteration of the genetic algorithm corresponds to an iteration of a map G, is a discrete dynamical system. The map G is a composition of a selection operator and a mixing operator, where the latter models the effects of both mutation and crossover. This paper shows that for a typical mixing operator, the fixed point set of G is finite. That is, an arbitrarily small perturbation of the mixing operator will result in a map G with finitely many fixed points. Further, any sufficiently small perturbation of the mixing operator preserves the finiteness of the fixed point set.

Keywords: Genetic algorithm, generic property, fixed point, transverse, transversality

1 Introduction

In this paper we study a dynamical systems model of the genetic algorithm (GA). This model was introduced by Vose [1] and is further extended in [2] and [3]. The dynamical systems model of the genetic algorithm provides an attractive mathematical framework for investigating the properties of GAs.

A practical implementation of the genetic algorithm seeks solutions in a finite search space which we denote $\Omega = \{1, 2, \ldots, n\}$. Each element of Ω can be thought of as a "species" with a given fitness value; the goal of the algorithm is to maximize the fitness. Usually there are multiple species with high fitness value and n is large. In order to avoid local maxima the GA algorithm uses mutation and crossover operations to maintain diversity in the pool of r individuals, representing the n species. The infinite population model considers an infinite population of individuals represented by the probability distribution over Ω,

$$P = (P_1, P_2, \ldots, P_n)$$

where P_i is the proportion of the i-th species in the population. An update of the genetic algorithm consists of mutation, selection and crossover steps and is represented in the infinite population model as an iteration of a fixed function G.

C.R. Stephens et al. (Eds.): FOGA 2007, LNCS 4436, pp. 97–109, 2007.

Although the precise correspondence between behavior of such infinite population genetic algorithm and the behavior of the GA for finite population sizes has not been established in detail, the infinite population model has the advantage of being a well defined dynamical system. Therefore, the techniques of dynamical systems theory can be used to formulate and answer some fundamental questions about the GA.

The fixed points are fundamental objects of interest in our study, because in realistic implementations of genetic algorithms one usually observes convergence to a fixed point. Convergence to a unique fixed point is automatic when the quadratic map is a contraction. However, this paper is about a more subtle situation when the genetic algorithm is not a contraction. Wright and Bidwell [4] found examples of genetic algorithms with stable period two points, that is, examples of GAs for which not all solutions converge to a fixed point. These examples, however, were constructed using mutation and crossover operators which do not correspond to operators used in practice. In spite of this discovery, it is clear that one cannot expect to prove convergence to a fixed point for all genetic algorithms. Thus, the important problem is to carefully define the largest possible class of realistic genetic algorithms for which all solutions do converge to a fixed point. This problem is still open.

Wright and Vose [5] considered a class of mappings G that were a composition of a fixed mutation and crossover maps, and a proportional selection map. The set of fitness functions that correspond to the proportional selection was parameterized by the positive orthant. They have shown that for an open and dense set of such fitness functions, the corresponding map G has finitely many fixed points.

In this contribution we will take a different path. We consider a class of mappings $G = M \circ F$ where F is an arbitrary, but fixed, selection map and M is a mixing map from a class $\mathcal{M}$ described in Definition 2. This class is broad enough to include all mixing maps formed by a composition of the mutation and crossover maps described in monographs by Reeves and Rowe [2] and Vose [3]. We show that for a typical (i.e. open and dense) set of mixing maps, the corresponding map G has finitely many fixed points.

Theorem 1. *Let $G = M \circ F$ be a composition of a selection map F and a mixing operator M. Then for a typical mixing operator $M \in \mathcal{M}$, G has finitely many fixed points.*

The main tool in the proof of this result is the powerful notion of transversality. This idea has been successfully used in differential topology and dynamical systems for over forty years. Before we go into mathematical details, we wish to illustrate this notion on some simple examples. Consider a scalar function $f(x,a) := x^2 + a$, that depends on a real parameter a, and let W be the $x-$axis in the plane $\mathbb{R}^2$, see Figure 1. Notice that for all parameters $a \neq 0$ the graph of $f(x,a)$ either does not intersect W at all (for $a > 0$) or it intersects W in two points (for $a < 0$). Both of these situations are stable under small change in the parameter a. That is, if there are two intersections for $a = a_0$, then for all a_1 sufficiently close to a_0 the graph of $f(x,a_1)$ also intersects W in two points. A

Fig. 1. An exceptional value of a corresponds to non-transverse intersection of the graph of f and W (left figure); typical value of a corresponds to transverse intersection

similar statement is true for no intersection. We observe that the set of these "stable" values of a (i.e. $a \neq 0$) is open and dense in the set of all a. The value $a = 0$ is exceptional since arbitrary small change in a changes the number of intersections with W.

We can characterize geometrically the "stable" values of a by observing that the sum of the tangent space to the graph of f and the tangent space to W at all points x of intersection generates the tangent space to $\mathbb{R}^2$ at x. This tangent space is again $\mathbb{R}^2$. In this case we will say that f *is transversal to* W. For the exceptional value $a = 0$ these two tangent spaces at the point of intersection $x = 0$ coincide and therefore their sum is a strict subspace of the tangent space of $\mathbb{R}^2$ at 0. This is a just a sophisticated way of saying that the graph of f and W "cross" for the "stable" values of a and only "touch" for the exceptional values of a. Notice that we have to define "crossing" loosely since empty intersection of the graph of f and W counts as "crossing".

The fundamental mathematical result we will use states that, under certain assumptions, the set of parameter values for which a general parameterized function intersects a given manifold W transversally, is both open and dense in the set of all parameters.

In proving our result, the essential step is to transform the problem of finiteness of the fixed point set to a transversality problem. We will set $f(x, \mathcal{M}) := M(F(x)) - x = G(x) - x$, where instead of a scalar parameter a we have multidimensional parameter space $\mathcal{M}$ that characterizes the set of all admissible mixing operators. If we let $W = \{0\}$, then the intersections of the graph of f and W correspond precisely to the fixed points of G. To prove finiteness of the set of fixed points, we evoke another general transversality theorem (Theorem 3) which states that this set has finitely many components for every M for which we have transversality.

The paper is organized as follows. In Sect. 2 we carefully define the infinite population model, GA map and the set of admissible mixing operators $\mathcal{M}$. In Sect. 3 we review transversality and provide the necessary background. In Sect. 4 we prove the main result and conclude in Sect. 5.

2   The Infinite Population Genetic Algorithm

The genetic algorithm searches for solutions in the search space $\Omega = \{1, 2, \ldots n\}$; each element of Ω can be thought of as a type of individual. We consider a total population of size r with $r \ll n$. We represent such a population as an *incidence vector*:

$$v = (v_1, v_2, \ldots, v_n)^T$$

where v_i is the number of times the individual of type i appears in the population. It follows that $\sum_i v_i = r$. We also identify a population with the *population incidence vector*

$$p = (p_1, p_2, \ldots, p_n)^T$$

where $p_i = \frac{v_i}{r}$ is the proportion of the i-th individual in the population. The vector p can be viewed as a probability distribution over Ω. In this representation, the iterations of the genetic algorithm yield a sequence of vectors $p \in \Lambda^r$ where

$$\Lambda^r := \{(p_1, p_2, \ldots, p_n)^T \in \mathbb{R}^n \mid p_i = \frac{v_i}{r} \text{ and } v_i \in \{0, \ldots, r\} \text{ for all } i \in \{1, \ldots, n\}$$

$$\text{with } \sum_i v_i = r\} \, . \tag{1}$$

We define

$$\Lambda := \{x \in \mathbb{R}^n \mid \sum x_i = 1 \text{ and } x_i \geq 0 \text{ for all } i \in \{1, \ldots, n\}\} \, .$$

Note that $\Lambda^r \subset \Lambda \subset \mathbb{R}^n$, where Λ is the unit simplex in $\mathbb{R}^n$. Not every point $x \in \Lambda$ corresponds to a population incidence vector $p \in \Lambda^r$, with fixed population size r, since p has non-negative rational entries with denominator r. However, as the population size r gets arbitrarily large, Λ^r "becomes dense" in Λ, that is, $\cup_{r \geq N} \Lambda^r$ is dense in Λ for all N. Thus Λ may be viewed as a set of admissible states for infinite populations. We will use p to denote an arbitrary point in Λ^r and x to denote an arbitrary point in Λ. Thus p always represents a population incidence vector in a finite population and x the corresponding quantity in infinite population, which is the probability distribution over Ω. Unless otherwise indicated, $x \in \Lambda$ is a column vector.

The utility of the infinite population model and its relationship with the finite population models is often discussed. Our result in Theorem 1 implies that the equilibria of the finite, but large, population model are well separated. This separation and the number of the fixed points can be precisely linked to the separation and number of fixed points in the infinite population model. To see this let p_j^r denote a fixed point of the finite population GA of size r for $j = 1, \ldots, k$. We let p_j^∞, $j = 1, \ldots, k$, be the finite number of fixed points for the infinite population model. Assuming that all fixed points are hyperbolic, let $3\delta := \min_{i,j} \|p_j^\infty - p_i^\infty\|$. By continuity, for $r > N_1$ there are a finite number of fixed points $p_j^r, j = 1, \ldots, k$, such that

$$\|p_j^r - p_j^\infty\| < \delta.$$

Then for all $r > N_1$

$$||p_j^\infty - p_i^\infty|| < ||p_j^r - p_j^\infty|| + ||p_j^r - p_i^r|| + ||p_i^r - p_i^\infty|| < 2\delta + ||p_j^r - p_i^r||$$

and since

$$3\delta < ||p_j^\infty - p_i^\infty||$$

we have

$$||p_j^r - p_i^r|| > 3\delta - 2\delta = \delta. \tag{2}$$

Select N_2 such that $1/N_2 << \delta$. Then for all $r > N_2$ the distance δ between the fixed points of the finite population model is much larger than the distance between potential populations.

Let $G(x)$ represent the action of the genetic algorithm on $x \in \Lambda$, and assume that $G : \Lambda \to \Lambda$ is a differentiable map ([3]). The map G is a composition of three maps: selection, mutation, and crossover. We will now describe each of these in turn.

We let $F : \Lambda \to \Lambda$ represent the selection operator. The i-th component, $F_i(x)$, represents the probability that an individual of type i will result if selection is applied to $x \in \Lambda$. As an example, consider proportional selection where the probability of an individual $k \in \Omega$ being selected is

$$Pr[k|x] = \frac{x_k f_k}{\sum_{j \in \Omega} x_j f_j} ,$$

where $x \in \Lambda$ is the population incidence vector, and f_k, the k-th entry of the vector f, is the fitness of $k \in \Omega$. Define $diag(f)$ as the diagonal matrix with entries from f along the diagonal and zeros elsewhere. Then, for $F : \Lambda \to \Lambda$, proportional selection is defined as

$$F(x) = \frac{diag(f)x}{f^T x} .$$

We restrict our choice of selection operators, F, to those which are $\mathcal{C}^1$, that is, selection operators with continuous derivatives.

We let $U : \Lambda \to \Lambda$ represent mutation. Here U is an $n \times n$ real valued matrix with ij-th entry $u_{ij} > 0$ for all i, j, and where U_{ij} represents the probability that item $j \in \Omega$ mutates into $i \in \Omega$. That is, $(Ux)_k := \sum_i u_{ki} x_i$ is the probability an individual of type k will result after applying mutation to population x.

Let crossover, $C : \Lambda \to \Lambda$, be defined by

$$C(x) = (x^T C_1 x, \ldots, x^T C_n x)$$

for $x \in \Lambda$, where $C_1, \ldots, C_n$ is a sequence of symmetric non-negative $n \times n$ real valued matrices. Here $C_k(x)$ represents the probability that an individual k is created by applying crossover to population x.

Definition 1. *Let $Mat_n(\mathbb{R})$ represent the set of $n \times n$ matrices with real valued entries. An operator $A : \mathbb{R}^n \to \mathbb{R}^n$ is* quadratic *if there exist matrices $A_1, A_2, \ldots, A_n \in Mat_n(\mathbb{R})$ such that $A(x) = (x^T A_1 x, \ldots, x^T A_n x)$. We denote a quadratic operator with its corresponding matrices as $A = (A_1, \ldots, A_n)$.*

Thus, the crossover operator, $C = (C_1, \ldots, C_n)$, is a quadratic operator ([6]).

We combine mutation and crossover to obtain the mixing operator $M := C \circ U$. The k-th component of the mixing operator

$$M_k(x) = x^T (U^T C_k U) x$$

represents the probability that an individual of type k will result after applying mutation and crossover to population x. Since C_k is symmetric, M_k is symmetric. Further, since C_k is non-negative and U is positive for all k, M_k is also positive for all k. Additionally, it is easy to see check that since $\sum_{k=1}^{n} [M_k]_{ij} = 1$, $M : \Lambda \to \Lambda$, and mixing is also a quadratic operator ([6]). Here $[M_k]_{ij}$ denotes the ij-th entry of the matrix M_k. This motivates the following general definition of a mixing operator.

Definition 2. *Let $Mat_n(\mathbb{R})$ represent the set of $n \times n$ matrices with real valued entries. We call a quadratic operator, $M = (M_1, \ldots, M_n)$, a* mixing operator *if the following properties hold:*

1. *$M_k \in Mat_n(\mathbb{R})$ is symmetric for all $k = 1, \ldots, n$;*
2. *$(M_k)_{ij} > 0$ for all $i, j \in \{1, \ldots, n\}$, and for all $k = 1, \ldots, n$;*
3. *$\sum_{k=1}^{n} [M_k]_{ij} = 1$ for all $j = 1, \ldots, n$ and $i = 1, \ldots, n$.*

Let $\mathcal{M}$ be the set of quadratic operators M satisfying (1)-(3). It is easy to see that (3) implies that $M \in \mathcal{M}$ maps Λ to Λ. We define a norm, $|| \cdot ||$, on $\mathcal{M}$ by considering for $M \in \mathcal{M}$, $M \in \mathbb{R}^{n^3}$, and using the Euclidean norm. For an alternative norm on the set of quadratic operators, see [7].

Definition 3. *We set*

$$G := M \circ F, \text{ for } M \in \mathcal{M} \tag{3}$$

to be the complete operator for the genetic algorithm, or a GA map.

We extend the domain of definition of F to the positive cone in $\mathbb{R}^n$, denoted $\mathbb{R}^{n+}$. The extension of F is denoted $\tilde{F}$ and is defined by

$$\tilde{F}(u) := F \left(\frac{u}{\sum_i u_i} \right).$$

Thus $\tilde{F}|_\Lambda = F$, and for $x \in \Lambda$, $D\tilde{F}(x)|_\Lambda = DF(x)$, the Jacobian of F. Let $\mathbb{R}_0^n := \{x \in \mathbb{R}^n | \sum_i x_i = 0\}$. Since $\mathbb{R}_0^n$ is the tangent space to Λ at x and $F(\Lambda) \subset \Lambda$, $D\tilde{F}(\mathbb{R}_0^n) \subseteq \mathbb{R}_0^n$. Because $\tilde{F} : \mathbb{R}^{n+} \to \Lambda$, it is also clear that the map G is extended to a map $\tilde{G} : \mathbb{R}^{n+} \to \Lambda$ and the preceding remarks apply to $\tilde{G}$ as well. In order to simplify the notation we will use symbols F and G for these extended functions.

Definition 4. *If $f(x) = x$, a point x is called a* fixed point *of f.*

Definition 5. *A property is* typical, *or* generic, *in a set S, if it holds for an open and dense set of parameter values in S.*

3   Transversality: Background and Terminology

This section provides the reader with the necessary background in differential topology. The material provided below follows [8]. Other references include [9], [10] and [11].

Let X, Y be n and m dimensional manifolds, respectively. For $x \in X$, we let $T_x X$ denote the tangent space to X at x. For a differentiable map $f : X \to Y$, we let $df_x : T_x X \to T_{f(x)} Y$ denote the derivative of the map f at the point x. In the special case that $T_x X = \mathbb{R}^n$ and $T_{f(x)} Y = \mathbb{R}^m$, we note that $df_x : \mathbb{R}^n \to \mathbb{R}^m$ is given by the Jacobian matrix $Df(x)$ ([12]).

The following notation is adopted from [8]. Let $\mathcal{A}, X, Y$ be $\mathcal{C}^r$ manifolds. Let $\mathcal{C}^r(X, Y)$ be the set of $\mathcal{C}^r$ maps from X to Y.

Definition 6 (Restricted definition from [8] for finite vector spaces). *Let X and Y be $\mathcal{C}^1$ manifolds, $f : X \to Y$ a $\mathcal{C}^1$ map, and $W \subset Y$ a submanifold. We say that f is transversal to W at a point $x \in X$, in symbols: $f \pitchfork_x W$, if, where $y = f(x)$, either $y \notin W$ or $y \in W$ and*

$$(T_x f)(T_x X) + T_y W = T_y Y .$$

We say f is transversal to W, in symbols: $f \pitchfork W$, if and only $f \pitchfork_x W$ for every $x \in X$.

We reformulate the example in Figure 1 in terms of Definition 6. Let $X := [-1, 1], Y := \mathbb{R}^2$ and let $W := \{(x, y) \in \mathbb{R}^2 \mid y = 0\}$ be the x-axis. We define the the family of maps $f_a : [-1, 1] \to \mathbb{R}^2$ by $f_a(t) := (t, t^2 + a)$. Then the map f_{-1} is transversal to W (Figure 1 right), but f_0 is not transversal to W (Figure 1 left).

Theorem 2 (Transversal Density Theorem, [8]). *Let $\mathcal{A}, X, Y$ be $\mathcal{C}^r$ manifolds and $W \subset Y$ a submanifold. Let $\rho_a : X \to Y$ be a family of maps such that the correspondence $ev(a, x) := \rho_a(x)$ is $\mathcal{C}^r$. Define $\mathcal{A}_W \subset \mathcal{A}$ by*

$$\mathcal{A}_W = \{a \in \mathcal{A} | \rho_a \pitchfork W\} .$$

Assume that

1. *X has finite dimension m and W has finite codimension q in Y;*
2. *$r > \max(0, m - q)$;*
3. *$ev(a, x) \pitchfork W$.*

Then $\mathcal{A}_W$ is residual (and hence dense) in $\mathcal{A}$.

In our example, the set $\mathcal{A}_W := \{a \neq 0\}$ and the set $\mathcal{A} = \mathbb{R}$. Density of $\mathcal{A}_W$ in $\mathcal{A}$ means that in an arbitrary neighborhood of the value $a = 0$, there is a value of the set $\mathcal{A}_W$.

Theorem 3 (Corollary 17.2, [8]). *Let X, Y be $\mathcal{C}^r$ manifolds ($r \geq 1$), $f : X \to Y$ a $\mathcal{C}^r$ map, $W \subset Y$ a $\mathcal{C}^r$ submanifold. Then if $f \pitchfork W$:*

1. *W and $f^{-1}(W)$ have the same codimension;*
2. *If W is closed and X is compact, $f^{-1}(\{W\})$ has only finitely many connected components.*

As an example, let $f_a : [-1, 1] \rightarrow \mathbb{R}^2$ be as defined above with $a \neq 0$. Since $W = \{(x, y) \in \mathbb{R}^2 | y = 0\}$, the set $f_a^{-1}(W)$ consists of finitely many points. In fact, this set consists of two points when $a < 0$ and no points when $a > 0$. Each of these points is a zero-dimensional submanifold of the dimension 1 interval $[-1, 1]$. Thus, $f_a^{-1}(W)$ has codimension 1. Since W is a 1-dimensional submanifold of $\mathbb{R}^2$, it also has codimension 1. Note that in our example, $f_0^{-1}(W) = 0$ and is a point. However, we can modify the family f_a in such a way that $f_0(-\epsilon, \epsilon) = 0$ and 0 is still the only value of a where we lack hyperbolicity, see Figure 2. This example shows that hyperbolicity is necessary for the conclusion of the Theorem 3.

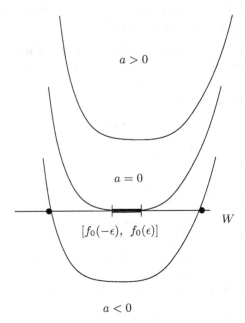

Fig. 2. A modified example from Figure 1. The map f_0 maps a subinterval $(-\epsilon, \epsilon)$ into W.

Observe that the compactness of X is also necessary. Consider the function $g : \mathbb{R} \rightarrow \mathbb{R}^2$ defined by $g(t) = (t, \sin t)$. Then $g^{-1}(0) = k\pi$ has infinitely many components in $X = \mathbb{R}$.

Theorem 4 (Openness of Transversal Intersection, [8]). *Let $\mathcal{A}, X, Y$ be C^1 manifolds with X finite dimensional, $W \subset Y$ a closed C^1 submanifold, $K \subset X$ a compact subset of X, and $\rho_a : X \rightarrow Y$ be a family of maps such that the correspondence $ev(a, x) = \rho_a(x)$ is C^r. Then the subset $\mathcal{A}_{KW} \subset \mathcal{A}$ defined by $\mathcal{A}_{KW} = \{a \in \mathcal{A} | \rho_a \pitchfork_x W \text{ for } x \in K\}$ is open.*

In our example, $K = X = [-1, 1]$ is compact and $\mathcal{A}_{KW} = \mathcal{A}_W := \{a \neq 0\}$. The set $\mathcal{A}_{KW}$ is open in $\mathcal{A} = \mathbb{R}$ if a small enough neighborhood, $N_\epsilon(a)$, of the value $a \in \mathcal{A}_{KW}$, has $N_\epsilon(a) \subset \mathcal{A}$. That is, for any $a \neq 0$, there is a small interval $(a - \epsilon, a + \epsilon) \subset \mathcal{A}$.

4   Proof of Main Results

Let $\mathcal{A} = \mathcal{M}$, where $\mathcal{M}$ denotes the set of mixing operators given by Definition 2, and let $X = \Lambda \subset \mathbb{R}^n$ and $Y = \mathbb{R}_0^n$. For $M \in \mathcal{M}$, we define a family of maps $\rho_M(x) : \Lambda \to \mathbb{R}_0^n$ by $\rho_M(x) := (M \circ F - I)x$. Recall that $\mathbb{R}_0^n := \{x \in \mathbb{R}^n | \sum_i x_i = 0\}$. Note that because $F, M \in \mathcal{C}^1(\Lambda, \Lambda)$, we also have $\rho_M \in \mathcal{C}^1(\Lambda, \mathbb{R}_0^n)$.

Finally, we define $\mathrm{ev}_\rho(M, x) : \mathcal{M} \times \Lambda \to \mathbb{R}_0^n$ by $\mathrm{ev}_\rho(M, x) := \rho_M(x)$ for $M \in \mathcal{M}$ and $x \in \Lambda$. That is,

$$\mathrm{ev}_\rho(M, x) := \rho_M(x) = M(F(x)) - x \ .$$

Finally, note that since G, F are $\mathcal{C}^1$, the function ev_ρ is also $\mathcal{C}^1$.

Lemma 1. *For* $\mathrm{ev}_\rho(M, x) := M(F(x)) - x$, $\mathrm{rank}(d(\mathrm{ev}_\rho)) = n - 1$.

Proof. Note first that since $\Lambda \subset \mathbb{R}^{n+}$, the derivative $d\,\mathrm{ev}_\rho$ of ev_ρ, $d\,\mathrm{ev}_\rho = D\,\mathrm{ev}_\rho$, is a Jacobian of ev_ρ. Similarly, we note that

$$D(\mathrm{ev}_\rho|_{(\mathcal{M} \times \Lambda)}) = D\mathrm{ev}_\rho|_{T_{(P,y)}(\mathcal{M} \times \Lambda)} \ .$$

Because $T(\mathbb{R}_0^n) = \mathbb{R}_0^n$, and

$$T(\mathcal{M} \times \Lambda) = \{(P, y) | P = (P_1, \ldots, P_n) \text{ with } \sum_i P_i = 0 \text{ and } y \in \mathbb{R}^n\} \ ,$$

it suffices to show that the Jacobian $D\,\mathrm{ev}_\rho : T(\mathcal{M} \times \Lambda) \to \mathbb{R}_0^n$ is onto. Thus, for $(P, y) \in T(\mathcal{M} \times \Lambda)$, we calculate

$$D\,\mathrm{ev}_{(M,x)}(P, y) = \left[\frac{\partial \mathrm{ev}_\rho}{\partial M}, \frac{\partial \mathrm{ev}_\rho}{\partial x}\right]\begin{pmatrix} P \\ y \end{pmatrix} = \frac{\partial \mathrm{ev}_\rho}{\partial M}P + \frac{\partial \mathrm{ev}_\rho}{\partial x}y \ .$$

By a short computation we get

$$\frac{\partial \mathrm{ev}_\rho}{\partial M}P = PF(x) \ ,$$

and

$$\frac{\partial \mathrm{ev}_\rho}{\partial x}y = 2(MF(x))DF(x)y - y \ .$$

Finally, for any $z \in \mathbb{R}_0^n$, and $(M, x) \in \mathcal{M} \times \Lambda$, we show there exists $(P, y) \in T(\mathcal{M} \times \Lambda)$ such that

$$D\mathrm{ev}_{(M,x)}(P, y) = PF(x) + 2(MF(x))DF(x)y - y = z \ . \tag{4}$$

We start by choosing $y = 0 \in \mathbb{R}^n$. Now, by (4), it suffices to find $P = (P_1, \ldots, P_n)$ such that

$$D\mathrm{ev}_{(M,x)}(P, y) = PF(x) + 0 - 0 = z \ . \tag{5}$$

Because $F : \mathbb{R}^{n+} \to \Lambda$, we let $u = F(x) \in \Lambda$. By (5), we see that for fixed $u \in \Lambda$, we want $P = (P_1, \ldots, P_n)$ such that $Pu = (u^T P_1 u, \ldots, u^T P_n u) = z$. Clearly, for $i = 1, \ldots, n-1$, we can choose P_i such that $u^T P_i u = z_i$. Finally, because $z \in \mathbb{R}_0^n$, that is $\sum_i z_i = 0$, we see that $z_n = -\sum_i z_i$. Thus, for our choice of $P_1, \ldots, P_{n-1}$,

$$z_n = -\sum_i z_i = -\sum_i u^T P_i u = u^T (\sum_i P_i) u \; ,$$

and $P_n = -\sum_{i=1}^{n-1} P_i$. Because $P_n = -\sum_{i=1}^{n-1} P_i$, it is clear that $\sum_{i=1}^{n} P_i = 0$, and for this choice of $P = (P_1, \ldots, P_n)$ with $y = 0$ we have $Dev_{(M,x)}(P, y) = z$. Because z, M, x were arbitrary, we have shown that $Dev_\rho : T(\mathcal{M} \times \Lambda) \to \mathbb{R}_0^n$ is onto. That is, rank$(Dev_\rho) = n - 1$.   □

Lemma 2. *Let* $ev_\rho : \mathcal{M} \times \Lambda \to \mathbb{R}_0^n$, $x \in \Lambda$, *and* $M \in \mathcal{M}$. *Then* $ev_\rho \pitchfork \{0\}$.

Proof. Choose $W := \{0\}$. Since $T_0 W = \{0\}$, to prove transversality using Definition 6, we need to show that the image $(T_x ev_\rho)(T_x X) = T_0 Y = \mathbb{R}^{n-1}$. In other words, we need to show that $Dev_\rho(x)$ is surjective. By Lemma 1, rank$(Dev_\rho(x)) = n - 1$, and therefore $Dev_\rho(x)$ is surjective and $ev_\rho \pitchfork \{0\}$.   □

Proposition 1. *Let* $\mathcal{M}_{\{0\}} := \{M \in \mathcal{M} | \rho_M \pitchfork \{0\}\}$. *Then* $\mathcal{M}_{\{0\}}$ *is dense in* $\mathcal{M}$. *That is, the set of parameter values for which* ρ_M *is transversal to* $\{0\}$ *is dense in* $\mathcal{M}$.

Proof. We apply the Transversal Density Theorem: Theorem 2. We first note that by Lemma 2, $ev_\rho \pitchfork \{0\}$, and therefore condition (3) of Theorem 2 holds. We now verify the remaining conditions (1)-(2).

1. $X = \Lambda$ has finite dimension m and $W = \{0\}$ has finite codimension q in $Y = \mathbb{R}_0^n$. Because $X = \Lambda$, $m = n - 1 < \infty$. Clearly, the codimension of $\{0\}$ in $\mathbb{R}_0^n$ is $n - 1$. That is, $q = n - 1$.
2. $r > \max(0, m - q)$. Since $r = 1$, clearly $r > \max(0, 0) = 0$.   □

Proposition 2. *The set* $\mathcal{M}_{\{0\}}$ *is open in* $\mathcal{M}$.

Proof. We apply Theorem 4, and therefore start by verifying its hypothesis conditions. Clearly $\mathcal{M}, X, Y$ are C^1 manifolds. We take $K = X = \Lambda$, and thus K is a compact subset of the finite dimensional manifold X. Similarly $W = \{0\} \subset Y$ is closed. By the previous argument, the maps $\rho_M(x)$ are C^1. Thus, all hypothesis requirements have been met and by Theorem 4,

$$\mathcal{M}_{K\{0\}} = \{M \in \mathcal{M} | \rho_M \pitchfork_x \{0\} \text{ for } x \in K = X\}$$

is open in $\mathcal{M}$.   □

Proposition 3. *For generic* $M \in \mathcal{M}$,

1. $\rho_M \pitchfork \{0\}$. *That is, the set of parameter values for which* ρ_M *is transversal to* $\{0\}$ *is open and dense in* $\mathcal{M}$.

2. *The set of parameter values for which $\rho_M^{-1}(\{0\})$ has finitely many solutions is open and dense in $\mathcal{M}$.*

Proof. The proof of part (1) follows directly from Lemmas 1 and 2.

We now prove part (2). By part (1), the set of parameter values for which ρ_M is transversal to $\{0\}$ is open and dense in $\mathcal{M}$. Thus by Theorem 3, for this open and dense set, $\rho_M^{-1}(\{0\})$ has only finitely many connected components. We now show by contradiction that there are finitely many solutions to $\rho_M(x) = 0$ in Λ.

For $x \in \rho_M^{-1}(\{0\})$, let $M_x \subset \rho_M^{-1}(\{0\})$ denote the connected component with $x \in M_x$. Assume x is not isolated in $\rho_M^{-1}(\{0\})$. Then, there exists a sequence $\{x_n\} \subset M_x$ such that $x_n \to x$, and by choosing a subsequence $\{x_{n_k}\}$,

$$\lim_{n_k \to \infty} \frac{x_{n_k} - x}{\|x_{n_k} - x\|} = v, \tag{6}$$

where $v \in T_x(\rho_M^{-1}(\{0\}))$. Here $v \neq 0$ because the terms in the quotient are on the unit sphere. Since

$$M \circ F(x_{n_k}) - M \circ F(x) = D(M \circ F)(x) \cdot (x_{n_k} - x) + R \tag{7}$$

where R is a remainder, then

$$\lim_{n_k \to \infty} \frac{M \circ F(x_{n_k}) - M \circ F(x)}{\|x_{n_k} - x\|} = \lim_{n_k \to \infty} \frac{D(M \circ F)(x) \cdot (x_{n_k} - x) + R}{\|x_{n_k} - x\|}. \tag{8}$$

By (6) and (8), and because x_{n_k}, x are fixed points,

$$v = D(M \circ F)(x) \cdot \left(\lim_{n_k \to \infty} \frac{x_{n_k} - x}{\|x_{n_k} - x\|} \right) + \lim_{n_k \to \infty} \frac{R}{\|x_{n_k} - x\|}. \tag{9}$$

That is,

$$v = D(M \circ F)(x) \cdot v + 0, \tag{10}$$

and $v \neq 0$ is an eigenvector of $D(M \circ F)(x)$ with eigenvalue 1. However, since $x \in \rho_M^{-1}(\{0\}$, and ρ_M is transversal to $\{0\}$ at x, $D(M \circ F)(x) - I$ is a linear isomorphism on a finite vector space. Thus, for all $v \neq 0$, $(D(M \circ F)(x) - I)v \neq 0$ which is a contradiction. Thus, all the components of $\rho_M^{-1}(\{0\})$ only contain isolated points and each connected component of $\rho_M^{-1}(\{0\})$ is itself an isolated point. Since there are finitely many connected components of $\rho_M^{-1}(\{0\})$, there are finitely many solutions to $\rho_M(x) = 0$ for $x \in \Lambda$. $\qquad\square$

Proof of Theorem 1. By Lemma 3, for $M \in \mathcal{M}_{\{0\}} \subset \mathcal{M}$, $\rho_M(x) = 0$ has finitely many solutions in Λ. That is, for generic $M \in \mathcal{M}$,

$$\rho_M(x) = M(F(x)) - x$$

has finitely many solutions in Λ. Thus solutions to $\rho_M(x) = 0$ correspond to fixed points of $G = M \circ F$. $\qquad\square$

5   Conclusions

In this paper we have shown that given an arbitrary selection function and a typical mixing function, their composition has finitely many fixed points. This composition represents an infinite population model of a GA. Even though the correspondence between the infinite population model of a GA and the finite population models that are used by practitioners is not straightforward and likely depends on the details of that implementation, our result adds to the increasing body of evidence that the infinite population model can give qualitative insights into the functioning of the GA. Genericity of the finiteness of the fixed point set is expected for any reasonably rich model, consisting of iterations of a map on a compact space. Our results can be interpreted as showing that the GA map and the class of mixing operators constitute such model. That is, as analogously stated in [5], for a given mixing operator, unless proven otherwise, it is reasonable to assume that G has finitely many fixed points.

We note that the perturbation from an infinite to a large finite population model can be viewed as a small perturbation of the infinite population model. Furthermore, our result in Theorem 1 implies that the fixed points of the large population limit are separated at least by a constant distance, as shown by (2). For large enough population sizes, this constant distance is much larger than the resolution between neighboring states of the GA.

The additional contribution of our work is to once again (after [5]) bring the attention of the GA community to a set of powerful ideas from differential topology, centered around the notion of transversality. We believe that these ideas can be applied in different contexts to many problems in the study of genetic algorithms.

Acknowledgement. T. G. was partially supported by NSF/NIH grant 0443863, NIH-NCRR grant PR16445 and NSF-CRCNS grant 0515290.

References

1. Wright, A.H., Vose, M.D.: Stability of vertex fixed points and applications. In: Foundations of Genetic Algorithms 3, vol. 3, Morgan Kaufman Publishers, San Francisco (1995)
2. Reeves, C.R., Rowe, J.E.: Genetic Algorithms - Principles and Perspectives: A Guide to GA Theory. Kluwer Academic Publishers, Boston, MA (2003)
3. Vose, M.D.: The Simple Genetic Algorithm: Foundations and Theory. MIT Press, Cambridge (1999)
4. Wright, A.H., Bidwell, G.: A search for counterexamples to two conjecture on the simple genetic algorithm. In: Foundations of Genetic Algorithms 4, vol. 4, Morgan Kaufman Publishers, San Francisco (1997)
5. Wright, A.H., Vose, M.D.: Finiteness of the fixed point set for the simple genetic algorithm. Evolutionary Computation 3(4) (1995)
6. Rowe, J.E., Vose, M.D., Wright, A.H.: Group properties of crossover and mutation. Evolutionary Computation 10(2), 151–184 (2002)

7. Rowe, J.E.: A normed space of genetic operators with applications to scalability issues. Evolutionary Computation 9(1) (2001)
8. Abraham, R., Robbin, J.: Transversal Mappings and Flows. W. A. Benjamin, Inc. New York (1967)
9. Guillemin, V., Pollack, A.: Differential Topology. Prentice-Hall, Inc. Englewood Cliffs (1974)
10. Hirsch, M.: Differential Topology. Springer, Heidelberg (1976)
11. J.Jr., P., de Melo, W.: Geometric Theory of Dynamical Systems. Springer, Heidelberg (1982)
12. Jänich, K.: Vector Analysis. Springer, Heidelberg (2001)

Neighborhood Graphs and Symmetric Genetic Operators

Jonathan E. Rowe[1], Michael D. Vose[2], and Alden H. Wright[3]

[1] School of Computer Science
University of Birmingham
Birmingham B15 2TT
Great Britain
J.E.Rowe@cs.bham.ac.uk
[2] Computer Science Department
University of Tennessee
Knoxville, TN 37996
USA
vose@cs.utk.edu
[3] Dept. of Computer Science
University of Montana
Missoula, Montana 59812
USA
wright@cs.umt.edu

Abstract. In the case where the search space has a group structure, classical genetic operators (mutation and two-parent crossover) which respect the group action are completely characterized by formulas defining them in terms of the search space and its group operation. This provides a representation-free implementation for those operators, in the sense that the genotypic encoding of search space elements is irrelevant. The implementations are parameterized by distributions which may be chosen arbitrarily, and which are analogous to specifying distributions for mutation and crossover masks.

1 Introduction

This paper extends the theory developed in [RVW02, RVW04] concerning groups that act transitively on a search space. The special case where the search space itself has a group structure (so that it acts transitively on itself) is the primary focus.

One might legitimately wonder what mixing operators for classical Genetic Algorithms are possible (given reasonable restrictions).[1] We have answered such a question; the main results completely characterize classical genetic operators – mutation and two-parent crossover – which respect search space symmetries.

---

[1] Admittedly, such questions concerning *genetic algorithms* are *foundational* rather than applied, but this paper was intended for *"Foundations Of Genetic Algorithms"*...

C.R. Stephens et al. (Eds.): FOGA 2007, LNCS 4436, pp. 110–122, 2007.
© Springer-Verlag Berlin Heidelberg 2007

The paper begins with motivation, describing how a search space might come to have a group acting upon it. This is done by way of first discussing neighborhood structures and their symmetries, next considering how neighborhood structures naturally arise as a consequence of neighborhood operators, and then describing conditions under which neighborhood operators have symmetries which may be ascribed to the search space itself. Consequently, the particular group ascribed to the search space can vary with the chosen neighborhood structure; more generally, it may be chosen arbitrarily. In any case, a transitive group action eliminates bias in the sense that the search space is made to look the same from every point. Results proved in sections 2 and 3 are probably not new; they are included as part of the motivation leading up to section 4. Additional motivation is provided by the fact that other authors have also considered group structures on the search space [Wei91, Sta96, RHKS02].

In the case where the search space has a group structure, the classical genetic operators (mutation and two-parent crossover) can be completely characterized when they commute with the group action. Moreover, they have representation-free implementations; the genotypic encoding of search space elements is irrelevant. Their implementations are parameterized by distributions which may be chosen arbitrarily, and which are analogous to specifying distributions for mutation and crossover masks.

1.1 Notation

Suppose the finite search space Ω is enumerated as $\{\omega_0, \ldots, \omega_{n-1}\}$. Without loss of generality Ω may be regarded as $\{0, \ldots, n-1\}$ through the association $i \mapsto \omega_i$.

The notation [*expression*] denotes 1 if *expression* is true, and 0 otherwise.

To maintain continuity with the thread of most relevant results [Vos99b], [RVW02], [RVW04], the "twist" of the matrix A (see section 4.1) is denoted by A^*.[2] A similar comment is apropos to how group operations are denoted. In the special case where a group can be expressed as a nontrivial direct sum of normal subgroups, the best choice would indubitably be $\oplus$ so as to be consistent with [Vos99b, RVW04] which are most relevant. This paper, however, concerns the general case; it uses $\circ$ to denote the group operation, which is consistent with [RVW02].

2 Neighborhood Structures

Assume the finite search space Ω has a *neighborhood structure*: every $x \in \Omega$ has a set $N(x) \subset \Omega$ of neighbors. A neighborhood structure N is equivalent to a *neighborhood graph* which has Ω as vertex set and which contains directed edge $x \to y$ iff $y \in N(x)$.

---

[2] We apologize for using superscript asterisk to denote something other than Kleene Closure...

Definition 1. *A neighborhood structure on a finite search space Ω is a function $N : \Omega \to 2^{\Omega}$ which to each $x \in \Omega$ assigns a set $N(x) \subset \Omega$ of neighbors.*

Definition 2. *The neighborhood graph corresponding to a neighborhood structure N has the domain Ω of N as vertex set and contains directed edge $x \to y$ iff $y \in N(x)$.*

In practice, a neighborhood structure often has symmetries. An *automorphism* of N is a bijection $\pi : \Omega \to \Omega$ such that if y is a neighbor of x, then $\pi(y)$ is a neighbor of $\pi(x)$. Equivalently, it is an invertible map (permutation) on the vertices of the neighborhood graph which preserves edges.

Definition 3. *An automorphism of a neighborhood structure N is a bijection $\pi : \Omega \to \Omega$ such that if y is a neighbor of x, then $\pi(y)$ is a neighbor of $\pi(x)$.*

The set $\mathcal{A}_N$ of all such automorphisms is the *symmetry group* of N (it is a group under function composition since Ω is finite); elements of $\mathcal{A}_N$ are called *symmetries* (of N).

Definition 4. *The symmetry group of a neighborhood structure N is the set $\mathcal{A}_N$ of all automorphisms of N (it is a group under function composition); elements of $\mathcal{A}_N$ are called symmetries of N.*

As a consequence of preserving edges of the neighborhood graph, the symmetry group of N *commutes* with the neighborhood structure; for all $x \in \Omega$ and all $\pi \in \mathcal{A}_N$

$$\pi \circ N(x) = \{\pi(y) : y \in N(x)\}$$
$$= \{z : z \in N(\pi(x))\}$$
$$= N(\pi(x))$$

The theory developed in [RVW02, RVW04] concerns groups that act *transitively* on Ω : for every $x, y \in \Omega$ there exists a group element g such that $g(x) = y$. A direct consequence of $\mathcal{A}_N$ commuting with N is that a necessary (but not sufficient) condition for transitivity is that the neighborhood graph be regular (all vertices have the same degree).

Definition 5. *([Big71]) A group $(G, \circ)$ is said to act on Ω if its elements act as permutations (of Ω) such that for all $g, g' \in G$, and all $x \in \Omega$,*

$$(g \circ g')(x) = g(g'(x))$$
$$e(x) = x$$

where $e \in G$ denotes the identity element. Moreover, G acts transitively (on Ω) if for every $x, y \in \Omega$ there exists a group element g such that $g(x) = y$.

2.1   Neighborhood Operators

A neighborhood structure might arise from a collection of *neighborhood operators* on Ω which could be used, for example, by a search algorithm; operator ν assigns to each x some particular neighbor $\nu(x)$. A collection $\mathcal{O}$ of neighborhood operators generates a neighborhood structure,

$$N(x) = \{\nu(x) : \nu \in \mathcal{O}\}$$

Definition 6. *A neighborhood operator is a function $\nu : \Omega \to \Omega$ which to each $x \in \Omega$ assigns a neighbor $\nu(x)$. The neighborhood structure N generated by a collection $\mathcal{O}$ of neighborhood operators maps $x \in \Omega$ to the set $N(x) = \{\nu(x) : \nu \in \mathcal{O}\}$.*

Example 1. Let $\beta(n)$ denote the ℓ-bit binary expansion of the integer n, and let $\oplus$ denote bitwise exclusive-or. The Hamming neighborhood structure on the set $\mathcal{S}$ of binary strings of length ℓ is generated by

$$\mathcal{O} = \{\nu_k : \forall x \in \mathcal{S} . \nu_k(x) = \beta(2^k) \oplus x, \ 0 \leq k < \ell\}$$

Here the operators of $\mathcal{O}$ are self-inverse, thus $x \in \mathcal{N}(y) \Longleftrightarrow y \in \mathcal{N}(x)$. The corresponding neighborhood graph is the Hamming cube.

If, as in the example 1, every element of $\mathcal{O}$ is invertible, then $\mathcal{O}$ is said to be invertible.

Definition 7. *A collection $\mathcal{O}$ of neighborhood operators is said to be invertible if every element of $\mathcal{O}$ is invertible.*

The neighborhood structure generated by a collection of neighborhood operators can be *connected*—meaning that the neighborhood graph is connected (there is a directed path from x to y for all $x, y \in \Omega$)—yet its symmetry group can fail to be transitive (see example 2 below).

Definition 8. *A neighborhood structure N is said to be connected if there is a directed path from x to y for all $x, y \in \Omega$.*

Example 2. Let $\mathcal{O}$ be the set of permutations $\{(120)(453), (210)(543), (012345), (543210)\}$ (in cycle notation) of $\Omega = \{0, 1, 2, 3, 4, 5\}$. The collection $\mathcal{O}$ of neighborhood operators is invertible (not because $\mathcal{O}$ is closed under inverse; what matters is that its elements are invertible). The neighborhood structure N generated by $\mathcal{O}$ is connected. The symmetry group of N is not transitive because the neighborhood graph is not regular (bi-directional edges are shown without arrow heads below)

Theorem 1. *Let N be the neighborhood structure generated by a collection $\mathcal{O}$ of neighborhood operators. The set G of all bijections of Ω which commute with all elements of $\mathcal{O}$ is a subgroup of $\mathcal{A}_N$.*

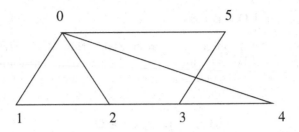

Fig. 1. Neighborhood graph of N

Proof. If $g, h \in G$ and $a \in \mathcal{O}$, then $g \circ h \circ a = g \circ a \circ h = a \circ g \circ h$, so $g \circ h \in G$. Moreover, $g \circ a = a \circ g$ implies $a \circ g^{-1} = g^{-1} \circ a$, and so $g \in G \Rightarrow g^{-1} \in G$. Hence G is a group.

Now suppose $y = a(x)$ for some $x, y \in \Omega$ and $a \in \mathcal{O}$, so that (x, y) is an edge of the neighborhood graph. If $g \in G$, then $g(y) = g \circ a(x) = a \circ g(x)$ so that $(g(x), g(y))$ is also an edge. Therefore $g \in \mathcal{A}_N$. □

Let N be the neighborhood structure generated by a collection $\mathcal{O}$ of neighborhood operators. Edge $x \to y$ in its neighborhood graph is said to have *color* ν if $\nu(x) = y$ and $\nu \in \mathcal{O}$ (an edge may have several colors). The group G (corresponding to $\mathcal{O}$ as in Theorem 1) not only preserves edges, it also preserves colors,

$$\nu(x) = y \iff g \circ \nu(x) = g(y) \iff \nu \circ g(x) = g(y)$$

The (edge-colored) neighborhood graph (of N) is said to correspond to $\mathcal{O}$. The group G is called the *symmetry group of $\mathcal{O}$*; its elements are called *symmetries* (of $\mathcal{O}$).

Definition 9. *The group of all bijections of Ω which commute with all elements of $\mathcal{O}$ is called the symmetry group of $\mathcal{O}$; its elements are called symmetries (of $\mathcal{O}$). The (edge-colored) neighborhood graph corresponding to the neighborhood structure generated by $\mathcal{O}$ is said to correspond to $\mathcal{O}$; edge $x \to y$ has color ν if $\nu(x) = y$ (for $\nu \in \mathcal{O}$; an edge may have several colors).*

Note that (by Theorem 1) the symmetry group of $\mathcal{O}$ is a subgroup of the symmetry group of the neighborhood graph corresponding to $\mathcal{O}$; the former preserves edges *and* color, whereas the latter need preserve only edges. For instance, the permutation (02)(35) is a symmetry of Figure 1 (i.e., a symmetry of N) but it is not a symmetry of the collection of neighborhood operators in example 1.

The next section deals with symmetries of $\mathcal{O}$.

3 Transitive Automorphism Groups

When using neighborhood operators as the basis of a search algorithm, the corresponding neighborhood graph is typically connected. As illustrated by example 2, connectivity is insufficient for the symmetry group of $\mathcal{O}$ to act transitively on Ω.

Nevertheless, when $\mathcal{O}$ acts on Ω the situation is quite different. If $\mathcal{O}$ is invertible, the group $\langle \mathcal{O} \rangle$ generated by $\mathcal{O}$ (under function composition) need not necessarily consist of symmetries of $\mathcal{O}$ (unless $\langle \mathcal{O} \rangle$ is Abelian), but it must act transitively; requiring that for all $x, y \in \Omega$ there exist $g \in \langle \mathcal{O} \rangle$ such that $g(x) = y$ is simply a restatement of connectivity.

Lemma 1. *Suppose the neighborhood graph corresponding to a collection $\mathcal{O}$ of neighborhood operators is connected. The only symmetry of $\mathcal{O}$ which has a fixed point is the identity (symmetry g has fixed point x iff $g(x) = x$).*

Proof. Let symmetry g and $x, y \in \Omega$ be such that $g(x) = x$. Since the neighborhood graph is connected, there exist $a_1, \ldots, a_k \in \mathcal{O}$ such that $y = a_1 \circ \cdots \circ a_k(x)$. Hence

$$g(y) = g \circ a_1 \circ \cdots \circ a_k(x) = a_1 \circ \cdots \circ a_k \circ g(x) = a_1 \circ \cdots \circ a_k(x) = y$$

Since y is arbitrary, g is the identity. □

Lemma 1, together with Theorem 17 from [RVW02], implies that if the neighborhood graph corresponding to $\mathcal{O}$ is connected, and if the symmetry group G of $\mathcal{O}$ acts transitivity, then there is a natural group structure isomorphic to G that can be ascribed to the search space itself. If the neighborhood graph corresponding to $\mathcal{O}$ is connected and if $\mathcal{O}$ is invertible, then the same conclusion holds with respect to $\langle \mathcal{O} \rangle$; there is a natural group structure isomorphic to $\langle \mathcal{O} \rangle$ that can be ascribed to Ω. These observations are recorded in the following theorem.

Theorem 2. *Suppose the neighborhood graph corresponding to a collection $\mathcal{O}$ of neighborhood operators is connected. If the symmetry group G of $\mathcal{O}$ acts transitively, then Ω has a group structure compatible with G (the search space can be given a group structure isomorphic to G such that the action of Ω on itself – via the group operation – is isomorphic to the action of G on Ω). If in addition $\mathcal{O}$ is invertible, then Ω has a group structure compatible with $\langle \mathcal{O} \rangle$.*

Proof. If G acts transitivity, then—*to use the language of* [RVW02]—it is a reduced group action on Ω (the only permutation fixing Ω is the identity), and Lemma 1 (above) implies that $\mathbf{Fix}(w) = \{0\}$, for all $w \in \Omega$ (only the identity fixes w); therefore Theorem 17 from [RVW02] applies to show that Ω has a group structure compatible with G.

If $\mathcal{O}$ is invertible, the comments preceding Lemma 1 imply the action of $\langle \mathcal{O} \rangle$ is reduced (the only permutation fixing Ω is the identity). Since $\nu(w) = w \implies \nu(g(w)) = g(\nu(w)) = g(w)$ for all $g \in G$, it follows that if ν fixes w then ν is the identity (since G acts transitively); hence Theorem 17 from [RVW02] applies to show that G has a group structure compatible with $\langle \mathcal{O} \rangle$. □

Suppose that one starts with a connected undirected neighborhood graph. For each edge (x, y) of this graph, define an automorphism $a_{(x,y)}$ of Ω by $a_{(x,y)}(x) = y$, $a_{(x,y)}(y) = x$, and $a_{(x,y)}(z) = z$ for all $z \neq x, y$. (The permutation $a_{(x,y)}$ is the transposition that is denoted by (x, y) in cycle notation.) Let $\mathcal{O}$ be the collection of these automorphisms. We will call the elements of $\mathcal{O}$ the edge transpositions of the graph. Then the graph of the neighborhood structure generated by $\mathcal{O}$ is the graph that we started with, except that loops have been added at each vertex. One might wonder if the symmetry group corresponding to this $\mathcal{O}$ would give Ω as a group structure as in theorem 2. The following small example shows that this is not necessarily the case.

Let $\Omega = \{0, 1, 2, 3\}$ and let $\mathcal{O} = \{(0, 1), (1, 2), (2, 3), (0, 3)\}$, where the elements of $\mathcal{O}$ are permutations written in cycle notation. The corresponding neighborhood graph is the square with loops at each vertex which is connected, so $\langle \mathcal{O} \rangle$ acts transitively on Ω. But since $\langle \mathcal{O} \rangle$ consists of all permutations of Ω, the symmetry group G of $\mathcal{O}$ is identity, and theorem 2 does not apply.

The example can be generalized to $\Omega = \{0, 1, \ldots, n - 1\}$ by defining $\mathcal{O}$ to be the set of edge transpositions of the edges of any connected graph whose vertices are Ω. We can show by induction on n that $\langle \mathcal{O} \rangle$ is the set S_n of all permutations of Ω. The base case is trivial. The induction hypothesis is that the edge transpositions of a connected graph on the vertices $\{0, 1, \ldots, n - 2\}$ generates the set S_{n-1} of permutations of $\{0, 1, \ldots, n-2\}$. There must be an edge $(i, n-1)$ of the graph that contains $n-1$. We need to show that $S_{n-1} \cup \{(i, n-1)\}$ generates S_n. Let σ be any permutation of $S_n \setminus S_{n-1}$, and let $j = \sigma(n-1)$. Note that $(i, j) \cdot (i, n - 1) \cdot (i, j) = (j, n - 1)$. Let $\tau = (j, n - 1) \cdot \sigma \in S_{n-1}$. Then $\sigma = (j, n - 1) \cdot \tau = (i, j) \cdot (i, n - 1) \cdot (i, j) \cdot \tau$.

In the case where the neighborhood operators mutually commute (which happens in example 1, for instance), then the following result holds.

Corollary 1. *Suppose the neighborhood graph corresponding to a collection $\mathcal{O}$ of neighborhood operators is connected. If $\mathcal{O}$ is invertible and if its elements commute, then its symmetry group is the group $\langle \mathcal{O} \rangle$ generated by $\mathcal{O}$ and Ω has a group structure compatible with $\langle \mathcal{O} \rangle$ (the search space can be given a group structure isomorphic to $\langle \mathcal{O} \rangle$ such that the action of Ω on itself – via the group operation – is isomorphic to the action of $\langle \mathcal{O} \rangle$ on Ω).*

Proof. If the bijections of $\mathcal{O}$ commute, then $\langle \mathcal{O} \rangle$ is a subgroup of the symmetry group G (of $\mathcal{O}$). Since the neighborhood graph is connected, the action of $\langle \mathcal{O} \rangle$ on Ω is transitive. Therefore, the action of G is also transitive and Theorem 2 applies. Note that $| G | = | \langle \mathcal{O} \rangle |$, since the search space has group structures isomorphic to G and to $\langle \mathcal{O} \rangle$. It follows that G is actually the same as $\langle \mathcal{O} \rangle$.   □

In the case corresponding to corollary 1, the group structure ascribed to Ω is commutative. Each element of Ω is identified with a list of those operators that must be applied in order to reach it from 0. In this case, the search space is *structural*: a situation which is dealt with in detail in [RVW04].

4   Implementation

Having described how a search space might come to have a group acting upon it, and may in fact have a group structure ascribed to it (so that it acts transitively on itself via the group operation), consider the issue of designing mutation and crossover operators for such search spaces.

For the remainder of the paper, assume that Ω does indeed have its own group structure (the group operation is denoted by $\circ$). Let $m(x)$ denote the result of mutating x, and let $c(\{x, y\})$ denote the result of crossing parents x and y.

The formal requirement for mutation and crossover to commute with the group $(\Omega, \circ)$ is cast within the framework of *Random Heuristic Search* [Vos99a], [Vos99b, RVW02]. A population is represented by a distribution $p \in \Lambda$, where

$$\Lambda = \{p \in \mathbf{R}^n : p_k \geq 0, \ \Sigma \, p_k = 1\}$$

and where p_i is the proportion of i in the population. An element a of the group $(\Omega, \circ)$ corresponds to a permutation matrix $\sigma_a : \Lambda \to \Lambda$ defined by

$$(\sigma_a)_{i,j} = [i = a \circ j]$$

Crossover acts on distribution p by mapping it to $\mathcal{C}(p)$ where

$$\mathcal{C}(p)_k = p^T M_k \, p$$

(superscript T denotes transpose) and where the matrix M_k is defined by

$$(M_k)_{i,j} = \text{Prob}\{c(\{i, k\}) = k\}$$

Crossover is said to *commute with* $(\Omega, \circ)$ if for all $p \in \Lambda$ and all $a \in \Omega$,

$$\mathcal{C}(\sigma_a p) = \sigma_a \mathcal{C}(p)$$

An advantage of considering genetic operators (crossover, mutation, selection) at the distribution level is that some analysis can proceed with differentiable objects, and, finite population information is preserved [Vos99b, RVW05, RVW06].

A similar situation holds for mutation; it has a corresponding operator $\mathcal{U}$ at the distribution level, and, it is said to commute with $(\Omega, \circ)$ if for all $p \in \Lambda$ and all $a \in \Omega$,

$$\mathcal{U}(\sigma_a p) = \sigma_a \mathcal{U}(p)$$

The definitions above – for what it means for crossover and mutation to commute with the group $(\Omega, \circ)$ – are given to provide context and to be technically accurate. However, they are not displayed for subsequent use in this paper; working definitions are instead provided by the following theorem.

Theorem 3. *Crossover and mutation commute with $(\Omega, \circ)$ if and only if for all $w, x, y, z \in \Omega$*

$$Prob\{w = m(x)\} = Prob\{z \circ w = m(z \circ x)\}$$
$$Prob\{w = c(\{x, y\})\} = Prob\{z \circ w = c(\{z \circ x, z \circ y\})\}$$

Proof. Theorem 3 is a rephrasing of Theorems 5 and 6 from [RVW02]. □

4.1   Binary Crossover

Let $\mathcal{B}$ be the set of maps from Ω to itself. Given a probability distribution χ over $\mathcal{B}$, form an offspring from parents $u, v \in \Omega$ by

1. choosing an element $b \in \mathcal{B}$ according to χ,
2. returning (with equal probability) an element from

$$\{v \circ b(v^{-1} \circ u), \ u \circ b(u^{-1} \circ v)\}$$

This crossover method is called the *canonical crossover scheme*.

Theorem 4. *The canonical crossover scheme commutes with* $(\Omega, \circ)$.

Proof. Define the function r by

$$r(u, v, w) = \sum_{b \in \mathcal{B}} \chi_b[v \circ b(v^{-1} \circ u) = w]$$

The probability that u and v cross to form w is

$$s(u, v, w) = \frac{r(u, v, w) + r(v, u, w)}{2}$$

For every $z \in \Omega$

$$r(z \circ u, z \circ v, z \circ w) = \sum_{b \in \mathcal{B}} \chi_b[z \circ v \circ b(v^{-1} \circ z^{-1} \circ z \circ u) = z \circ w]$$

$$= \sum_{b \in \mathcal{B}} \chi_b[v \circ b(v^{-1} \circ u) = w]$$

$$= r(u, v, w)$$

It follows from Theorem 3 that crossover commutes with $(\Omega, \circ)$.   □

Example 3. Let Ω be the set of length ℓ binary strings under the group operation $\oplus$ of bitwise exclusive-or (in particular, $v^{-1} = v$). For every $k \in \Omega$ define the map $b_k \in \mathcal{B}$ by

$$b_k(x) = k \otimes x$$

where $\otimes$ denotes bitwise and. Let χ be a probability distribution that only assigns non-zero weight to such maps, and thus can be thought of as a probability distribution over Ω. Therefore,

$$r(u, v, w) = \sum_{b \in \mathcal{B}} \chi_b[v \oplus b(v \oplus u) = w]$$

$$= \sum_{k \in \Omega} \chi_k[v \oplus (k \otimes (v \oplus u)) = w]$$

$$= \sum_{k \in \Omega} \chi_k[v \oplus (k \otimes v) \oplus (k \otimes u) = w]$$

$$= \sum_{k \in \Omega} \chi_k[((k \oplus \mathbf{1}) \otimes v) \oplus (k \otimes u) = w]$$

$$= \sum_{k \in \Omega} \chi_k[(\bar{k} \otimes v) \oplus (k \otimes u) = w]$$

where $\mathbf{1}$ is the string of all ones and $\bar{k}$ is the binary complement of k. It follows that the crossover scheme implements crossover by masks (see [Vos99b]).

According to Theorem 2 of [RVW02], every two-parent crossover commuting with $(\Omega, \circ)$ is completely determined by a mixing matrix M with the property that for all $i, j, k \in \Omega$,

$$\text{Prob}\{c(\{i, j\}) = k\} = M_{k^{-1} \circ i,\, k^{-1} \circ j}$$

Moreover, Theorem 19 of [RVW02][3] implies that $M = A^*$ for some row stochastic matrix A, where the *twist* A^* of A is defined by

$$A_{u,v}^* = A_{u^{-1} \circ v,\, u^{-1}}$$

In particular, $A = M^{**}$, since the twist operator has order three (i.e., $A = A^{***}$ for every matrix A). Another property of the twist is that $A^{*T} = A^{T**}$. Since mixing matrices are symmetric,

$$2M = A^* + A^{*T} = (A + A^{T*})^*$$

and therefore

$$\tfrac{1}{2}(A + A^{T*}) = M^{**}$$

Theorem 5. *Given any row stochastic matrix A, there exists a probability distribution α over $\mathcal{B}$ such that*

$$A_{i,j} = \sum_{f \in \mathcal{B}} \alpha_f [f(i) = j]$$

Moreover, given any such α, the formula above defines a row stochastic matrix.

Proof. Any matrix having the form above is row stochastic, since

$$\sum_j A_{i,j} = \sum_{f \in \mathcal{B}} \alpha_f \sum_j [f(i) = j] = \sum_{f \in \mathcal{B}} \alpha_f = 1$$

Conversely, if α is defined by

$$\alpha_f = \prod_{k \in \Omega} A_{k, f(k)}$$

then, using the identification $f \leftrightarrow \langle f_0, \ldots, f_{n-1} \rangle$ where $f_i = f(i)$,

$$\sum_{f \in \mathcal{B}} \alpha_f [f(i) = j] = \sum_{f_0 \in \Omega} \sum_{f_1 \in \Omega} \cdots \sum_{f_{n-1} \in \Omega} \left(\prod_{k \in \Omega} A_{k, f_k} \right) [f_i = j]$$

---

[3] The wording of Theorem 19 is directed towards *constructing* mixing matrices, hence the requirement to symmetrize (mixing matrices are symmetric). If one is representing a mixing matrix (see the discussion before Theorem 19), there is no need to symmetrize; mixing matrices are by definition symmetric.

$$= \sum_{f_0 \in \Omega} A_{0,f_0} \sum_{f_1 \in \Omega} A_{1,f_1} \cdots \sum_{f_{n-1} \in \Omega} A_{n-1,f_{n-1}}[f_i = j]$$

$$= \sum_{f_i \in \Omega} A_{i,f_i}[f_i = j]$$

$$= A_{i,j} \qquad\qquad \square$$

For every $f \in \mathcal{B}$, define the function $\widehat{f} \in \mathcal{B}$ by

$$\widehat{f}(x) = x \circ f(x^{-1})$$

The transformation $\widehat{} : \mathcal{B} \to \mathcal{B}$ is a bijection (it is self-inverse).

Theorem 6. *Every two-parent crossover commuting with $(\Omega, \circ)$ is an instance of the canonical crossover scheme.*

Proof. By what has been explained above, it suffices to choose the distribution χ in the canonical crossover scheme such that the resulting probability of obtaining k by crossing parents i and j (via the canonical scheme) is

$$M_{k^{-1}\circ i,\, k^{-1}\circ j} \qquad\qquad (1)$$

Define χ by

$$\chi_f = \tfrac{1}{2}(\alpha_f + \alpha_{\widehat{f}})$$

where α_f is the distribution referred to in Theorem 5 for $A = M^{**}$. Note that $\chi_f = \chi_{\widehat{f}}$, and the quantification $f \in \mathcal{B}$ is the same as $\widehat{f} \in \mathcal{B}$. The probability of obtaining k is

$$\tfrac{1}{2}\sum_{f \in \mathcal{B}} \chi_f[j \circ f(j^{-1} \circ i) = k] + \tfrac{1}{2}\sum_{f \in \mathcal{B}} \chi_{\widehat{f}}[i \circ \widehat{f}(i^{-1} \circ j) = k]$$

$$= \tfrac{1}{2}\sum_{f \in \mathcal{B}} \chi_f[j \circ f(j^{-1} \circ i) = k] + \tfrac{1}{2}\sum_{f \in \mathcal{B}} \chi_f[i \circ (i^{-1} \circ j) \circ f(j^{-1} \circ i) = k]$$

$$= \sum_{f \in \mathcal{B}} \chi_f[j \circ f(j^{-1} \circ i) = k]$$

$$= \tfrac{1}{2}\sum_{f \in \mathcal{B}} \alpha_f[f(j^{-1} \circ i) = j^{-1} \circ k] + \tfrac{1}{2}\sum_{f \in \mathcal{B}} \alpha_{\widehat{f}}[f(j^{-1} \circ i) = j^{-1} \circ k]$$

Setting $u = j^{-1} \circ i$, $v = j^{-1} \circ k$, and re-indexing the second sum in the last line above yields

$$\tfrac{1}{2}\sum_{f \in \mathcal{B}} \alpha_f[f(u) = v] + \tfrac{1}{2}\sum_{\widehat{f} \in \mathcal{B}} \alpha_f[\widehat{f}(u) = v]$$

$$= \tfrac{1}{2}A_{u,v} + \tfrac{1}{2}\sum_{f \in \mathcal{B}} \alpha_f[u \circ f(u^{-1}) = v]$$

$$= \tfrac{1}{2}A_{u,v} + \tfrac{1}{2}A_{u^{-1},\, u^{-1}\circ v}$$

$$= \tfrac{1}{2}(A + A^{T*})_{u,v}$$

$$= M^{**}_{u,v}$$

$$= M_{v^{-1},v^{-1}\circ u}$$

$$= M_{k^{-1}\circ j,\, k^{-1}\circ i}$$

This agrees with (1) since M is symmetric. $\square$

4.2   Mutation

Theorem 20 of [RVW02] and the Corollary following it describe how to implement mutation: given a probability distribution μ over Ω, mutate $j \in \Omega$ by

1. choosing an element $k \in \Omega$ according to μ,
2. returning the element $j \circ k$

According to the Theorem, all possible mutation operators which commute with $(\Omega, \circ)$ are of this form.

> **Example 4** Let Ω be the set of length ℓ binary strings under the group operation of bitwise exclusive-or. Then the mutation scheme above implements mutation by masks.

Mutation can be seen as a special case of crossover, in the sense that the resulting child will be the mutation of some parent. For each element $k \in \Omega$ define the map $b_k \in \mathcal{B}$ by

$$b_k(x) = x \circ k$$

Using these maps to implement crossover (nonzero probability is assigned by χ only to such maps), the set of possible children resulting from parents u, v and map b_k is

$$\{v \circ b_k(v^{-1} \circ u),\ u \circ b_k(u^{-1} \circ v)\} = \{u \circ k,\ v \circ k\}$$

Moreover, choosing χ to satisfy

$$\chi_{b_k} = \mu_k$$

arranges for the resulting child (of crossover) to not only be the mutation of a parent, but to occur according to the probabilities specified by μ.

5   Conclusion

This paper introduces neighborhood structures and their symmetries, and describes how neighborhood structures naturally arise as a consequence of neighborhood operators. Conditions are given under which neighborhood operators have symmetries which may be ascribed to the search space itself. This motivates—by providing a concrete example—the situation with which the paper is primarily concerned; the search space itself has a group structure.

Irrespective of how or why the search space may have a group structure, the main result is that those classical genetic operators (mutation and two-parent crossover) which respect the group action are completely characterized.

Formulas are given which define such genetic operators in terms of the search space and its group operation. This provides a representation-free implementation for those operators, in the sense that the genotypic encoding of search space elements is irrelevant. The implementations are parameterized by distributions which may be chosen arbitrarily, and which are analogous to specifying distributions for mutation and crossover masks (when specialized to a classical fixed-length binary GA, the standard crossover and mutation operators defined by masks result).

References

[Big71]   Biggs, N.L.: Finite groups of automorphisms. Cambridge University Press, Cambridge, UK (1971)

[RHKS02]   Rockmore, D., Hordijk, W., Kostelec, P., Stadler, P.: Fast fourier transforms for fitness landscapes. Applied and Computational Harmonic Analysis 12(1), 57–76 (2002)

[RVW02]   Rowe, J.E., Vose, M.D., Wright, A.H.: Group properties of crossover and mutation. Evolutionary Computation 10, 151–184 (2002)

[RVW04]   Rowe, J.E., Vose, M.D., Wright, A.H.: Structural search spaces and genetic operators. Evolutionary Computation 12, 461–494 (2004)

[RVW05]   Rowe, J.E., Vose, M.D., Wright, A.H.: Coarse graining selection and mutation. In: Wright, A.H., Vose, M.D., De Jong, K.A., Schmitt, L.M. (eds.) FOGA 2005. LNCS, vol. 3469, pp. 176–191. Springer, Heidelberg (2005)

[RVW06]   Rowe, J.E., Vose, M.D., Wright, A.H.: Differentiable coarse graining. Theoretical Computer Science 361, 111–129 (2006)

[Sta96]   Stadler, P.F.: Landscapes and their correlation functions. Journal of Mathematical Chemistry 20(1), 1–45 (1996)

[Vos99a]   Vose, M.D.: Random heuristic search. Theoretical Computer Science 229, 103–142 (1999)

[Vos99b]   Vose, M.D.: The Simple Genetic Algorithm: Foundations and Theory. MIT Press, Cambridge, MA (1999)

[Wei91]   Weinberger, E.D.: Fourier and taylor series on fitness landscapes. Biological Cybernetics 65(5), 321–330 (1991)

Decomposition of Fitness Functions in Random Heuristic Search

Yossi Borenstein and Riccardo Poli

Department of Computer Science, University of Essex, UK
{yboren,rpoli}@essex.ac.uk

Abstract. We show that a fitness function, when taken together with an algorithm, can be reformulated as a set of probability distributions. This set can, in some cases, be equivalently viewed as an information vector which gives ordering information about pairs of search points in the domain. Certain performance criteria definable over such an information vector can be learned by linear regression in such a way that extrapolations can sometimes be made: the regression can make performance predictions about functions it has not seen. In addition, the vector can be taken as a model of the fitness function and used to compute features of it like difficultly via vector calculations.

1   Introduction

Genetic algorithms (GAs) are problem independent heuristics which have been reported to perform relatively well on problems for which only partial knowledge is available. One of the main challenges of the field of evolutionary computation is to predict the behavior of GAs. In particular, the goal is to be able to classify problems as hard or easy according to the performance GA would be expected to have on such problems, before actually running the GA.

A first avenue in this direction was the Building Block (BB) hypothesis [8], which states that a GA tries to combine low, highly fit schemata. Following the BB hypothesis the notion of deception [8,6] isolation [9] and multimodality [19] have been defined. These were able to explain a variety of phenomena. Unfortunately, they did not succeed in giving a reliable measure of GA-hardness [10,11].

Given the connection between GAs and theoretical genetics, some attempts to explain the behavior of GAs were inspired by biology. For example, epistasis variance [4] and epistasis correlation [15] have been defined in order to estimate the hardness of a given real world problem. NK landscapes [2,13] use the same idea (epistasis) in order to create an artificial, arbitrary, landscape with a tunable degree of difficulty. These attempts, too, did not succeed in giving a full explanation of the behavior of a GA [11,16].

Finally, fitness distance correlation [12] tries to measure the intrinsic hardness of a landscape, independently of the search algorithm. Despite good success, fitness distance correlation is not able to predict performance in some scenarios [11].

The partial success of these approaches is not surprising. Several difficulties present themselves when developing a general theory that explains the behavior of a GA and is able to predict how it will perform on different problems.

C.R. Stephens et al. (Eds.): FOGA 2007, LNCS 4436, pp. 123–137, 2007.

A GA is actually a family of different algorithms. Given a problem, the GA designer first decides which representation (e.g. binary, multiary, permutation or real numbers) to use, then how to map the solution space into the search space, and finally which operator(s) (mutation, crossover) to use. Moreover, there are limited concrete guidelines on how to choose a representation and a genotype-phenotype mapping. Indeed this is a very difficult task. Different genotype-phenotype representations can completely change the difficulty of a problem [7]. There have been attempts to evolve the right representation [1] and there are some general design guidelines [14,18,7]. However, the reality is that the responsibility of coming up with good ingredients for a GA is still entirely on the GA designer.

In this paper we show that a fitness function, f, *when taken together with an algorithm*, can be reformulated as a vector of probability distributions, $\mathcal{P}_f$ (section 2). Following this decomposition, we are able to derive a first order approximation to the performance of Randomized Search Heuristics (RSHs). Using the GA as an example, we show that for small problems regression can be used to learn the coefficients of a linear model and predict the performance for unseen problems.

Using our construction, a measure of distance between two functions can be defined. We suggest to measure problem difficulty by measuring the distance between a function and the easiest possible function for an algorithm.

We measure the exploration ability of the algorithm as the expected entropy of the distribution that the selection mechanism defines (given a particular function) over the different search states. High values of entropy indicate that the expected performance of GA over the function is poor.

Due to the size of the sequence of probability distributions $\mathcal{P}_f$, the measurement suggested in section 2 cannot be used in practice. In section 3 we argue that $\mathcal{P}_f$ can (at least in some cases) be equivalently viewed as an information vector V, which gives ordering information about pairs of search points in the domain. While this is an approximation, it allows us to make concrete measurements and test our assumptions empirically. Section 3.1 suggests a first order approximation to the performance of RSHs on V. This is then tested empirically (section 4.1) on toy problems (V is much smaller than $\mathcal{P}_f$ however it is still too big). In section 3.3 we consider an indicator to the expected entropy defined in section 2.3. Using the first order approximation we show that the entropy may imply bounds on the expected performance. Finally, in section 3.2 we consider a vector V_{max} as the equivalent of the easiest fitness function. We argue that the distance between any other vector V and V_{max} can be used as an indication to problem difficulty. This is corroborated by experiments in section 4.2.

2   Algorithmic-Decomposition of the Fitness Function

Once all the parameters of a heuristic are chosen (i.e., representation, neighborhood structure, search operators etc.) the performance of the heuristic depends *solely* on the fitness function being optimized. In this section we show that it is possible to represent the function as k independent parameters (the actual number depends on the algorithm). We refer to this decomposition of the function as the *information content* of the function. We suggest to use a greedy criterion to measure the performance and derive two measurements of problem difficulty.

Let X denote a finite search space and $f : X \to Y$ a function. Let F denote all possible fitness functions. Using Vose's notation [21], RSHs can be thought of as an initial collection of elements $\Psi_k \in \Psi$ chosen from some search space X together with a transition rule τ which produces from the collection Ψ_k another collection Ψ_l. A collection of elements is a multiset of X. We use the term *search-state* to denote a particular collection. The set of all possible such collections, the state-space, is denoted by Ψ. Without loss of generality, we assume a notion of order in Ψ. The search is a sequence of iterations of τ: $\Psi_k \xrightarrow{\tau} \Psi_l \xrightarrow{\tau} \cdots$.

In reality, the transition rule $\tau(\Psi_i, f)$ if often a composition $\tau = \chi \circ \xi(\Psi_i, f)$ where $\xi(\Psi_i, f)$ denotes a *selection* operator, and χ can be thought of as the *navigation* operator. The selection phase identifies solutions with high fitness value, the navigation operator samples accordingly new solutions from X.

In order to make a clear distinction between the two operators (stages) it is useful to think of an output of the selection operator, a multiset, s, which represents a possible way to choose (or select) solutions from Ψ_i. For example, in GAs, given a particular population, Ψ_i, s is a possible mating pool. For a $(1+1)$ evolutionary strategy, s can be either the parent or the offspring. Given a state Ψ_i, we denote by S^i all possible ways of selecting points. That is, S^i is a set of multisets, each multiset, $s \in S^i$, corresponds to one possible way of selecting points.

The dependency of the performance of a search algorithm on f is reflected by a probability distribution, P_f^i, that the selection mechanism defines for each state over S^i. The particular multiset of solutions, s, that the algorithm selects, being the only argument for the navigation operator, *defines* (possibly, stochastically) the next state of the search. We define the *information content* of f as the set of all such distributions.

Definition 1. *The information content of a fitness function f is the set $\mathcal{P}_f = \{P_f^1, P_f^2, ..., P_f^n\}$ which gives for each state, Ψ_i the probability distribution P_f^i used in the selection phase.*

Usually, the algorithm does not define explicitly a probability distribution over S^i, rather, a distribution over single solutions from Ψ_i. For example, binary tournament selection defines the probability of selecting one of two possible solutions as follows:

$$\Pr_{tmt}\{x \mid \{x, y\}\} = \delta(f(x) > f(y)) + 0.5\delta(f(x) = f(y)) \tag{1}$$

where the function $\delta(\text{expr})$ returns 1 if expr is true, and 0 otherwise. This is used for a state (population) bigger than two points, by selecting, iteratively, uniformly at random, two points from Ψ_i and applying equation 1:

$$\Pr(x \mid \Psi_i, f) = \Pr\{x, x \mid \Psi_i\} + \sum_{x \neq y} \Pr\{x, y \mid \Psi_i\} \cdot \Pr_{tmt}\{x \mid \{x, y\}\} \tag{2}$$

Finally, P_f^i, the probability of selecting a particular multiset, s, is obtained as follows:

$$P_f^i(s \mid \Psi_i, f) = \prod_{x \in d} \Pr(x \mid \Psi_i, f). \qquad (3)$$

As previously mentioned, we focus only on the selection phase of the algorithm. Our analysis is done under the assumption that all the parameters of the algorithm (including the choice of representation or neighborhood structure) are defined. In this case, the performance of the algorithm depends on the fitness function alone, or using our formulation, the information content of the fitness function.

Since for each $f \in F$, P_f is properly defined, the set of all possible fitness functions F corresponds to a similar set, denoted by $\mathcal{P}_F$ of probability distributions. Let $\mathbf{P}(a, f)$ denote the performance of the algorithm, a, on the function f. We assume that $\mathbf{P}$ indicates the efficiency of the algorithm. It can, for example, denote the expected number of fitness evaluations required to find an optima point, but can be more general (e.g., expected number of generations for GA). We assume that the initial state is chosen uniformly at random and so, when the expectation is done over multiple runs, there is always some probability of the global optimum to be selected. Since we assume that a is fixed we consider the performance function as follows:

$$\mathbf{P} : \mathcal{P}_F \to \mathbb{R}$$

2.1   Greedy Criteria for Performance

Having no a priori assumption about $\mathcal{P}_f$ is equivalent to assuming that each $P_f^i \in \mathcal{P}_f$ is a uniform distribution over S^i. In that case, a distance between two states, $d(s, \Psi_j)$, can be defined as the expected first hitting time of a random walk starting from s and reaching Ψ_j.

Assuming that the algorithm tries at each step to minimize the distance to an optimum state a greedy criterion for the performance can be defined. Assuming that the optima is known, the efficiency of P_f^i can be measured as the expected distance at time step $t + 1$ to an optimum state – that is, $\sum P_f^i(s)d(s, \Psi_{opt})$ where Ψ_{opt} denotes a state which contains an optimum point. This suggests that the effect of each variable of the function $\mathbf{P}$ can be evaluated, to some extent, independently. Since, such an analysis can be done only if the global optimum is given, we write explicitly $\mathbf{P}(\mathcal{P}_f, x_{target})$ where $\mathcal{P}_f$ is the information content of the function f and x_{target} denotes the global optimum.

2.2   Distance and Performance

Following this line of reasoning, it is possible to define a distribution $\mathcal{P}_{f_{opt}}$ such that $d(s, \Psi_{opt})$ is minimized for each $P_{f_{opt}}^i$. The accuracy of any other distribution $\mathcal{P}_f$ can be evaluated by comparing how similar it is (e.g., using the expected Kullback-Leibler divergence) to $\mathcal{P}_{f_{opt}}$.

2.3   Number of Ties and Entropy

One thing that lead to poor performance in classical AI search is ties between competing alternative successor states [17]. This is because a searcher is then forced to either explores all of them or chooses one arbitrarily. The same happens in RSHs when a distribution P_f^i is uniform (or almost uniform). In that case, the algorithm chooses the next state uniformly at random. Clearly, if this is the case for all P_f^i's the algorithm performs random search.

A natural way of measuring the frequency of ties for stochastic algorithms is to consider the entropy of the distribution defined over all possible successors. Entropy measures how uniform a distribution is, that is, the larger the entropy the less informed the algorithm is.

Definition 2. *The entropy, $H(P_f^i)$, of the function f for the state Ψ_i, is defined as:*

$$H(P_f^i) = -\sum_{s \in S^i} P_f^i(s) \log P_f^i(s)$$

Depending on the function f, different search states, have different probabilities to occur in a run. For example, RSHs optimizing the Ridge function [11] are very likely to sample a state which contains the solution $x = 0^n$. So, we measure the overall effect of ties as the expected entropy:

$$E[H(f)] \equiv \sum_i Pr(\Psi_i) H(P_f^i)$$

where $Pr(\Psi_i)$ denotes the probability that the state Ψ_i will be sampled during a run.

When $E[H(f)]$ is maximal, either the function is almost flat, or the selection mechanism is random. The needle-in-a-haystack (NIAH) is a well known example for this scenario: the fitness of all the solutions but the global optimum equals 0. The performance of RSH on the NIAH is bounded from below by $(|X|+1)/2$ [5] which suggests that high expected entropy implies hardness. On the other hand, when $E[H(f)] = 0$ nothing can be said about the performance (which depends, in this case, solely on the relation between the search operators and the fitness function). The effect of intermediate values of entropy on performance is more difficult to asses. Presumably, the higher the entropy the closer the performance to that of a random search. Under the first order approximation considered in the next section – this is precisely the case.

It is worth mentioning that NP-hard problems exist where the expected entropy is maximal. SAT is an obvious example: from the black box perspective, a SAT instance is a variant of the NIAH with possibly more than one needle. Interestingly, while the fitness distribution of many MAXSAT problems is not flat, a NIAH-type of MAXSAT problem can be generated for n bit space using, for example, the following formula:

$$\bigcap_{i \geq 0} \left(x_i \bigcup_{j < i} \neg x_j \right)$$

where $x_i \in X$ are literals. This is definitely not a typical case, however, this is a clear example for important problems for which the entropy, from the *black box perspective*, is maximal.

3   Approximation of the Performance Function

The decomposition of f to $\mathcal{P}_f$ is precise, however, due to its size, it cannot be used in practice. In this section we define a simpler decomposition, in which every function f is associated with a vector V. In the reminder of the section we relate the material in section 2 to this simpler model. In particular, section 3.1 introduces a first order approximation to the performance. Section 3.2 defines the distance between a vector V to an optimal vector V_{max} as a predictive measure of problem hardness. Finally, in section 3.3 the entropy of f is approximated by using V.

The size of $|\Psi|$ and $|\Psi_i|$ for realistic search algorithms is usually bounded. For example, genetic algorithms typically use a population of fixed size, the size of a tabu-list is bounded and local-searchers often consider only two solutions at any given time. In order to have a more concrete formulation we will restrict our attention to algorithms that use a comparison of pairs of solutions in the search space. As illustrated in section 2, for these algorithms, equation (1) (the probability, given two solutions, that one is selected) is the only one that considers, explicitly the fitness function. Equation (2) and then (3) (the actual distribution, P_f^i) depend only on $\Pr_{tmt}$.

The codomain is the function $\Pr_{tmt}$ is the set of probabilities $\{0, 0.5, 1\}$. Equivalently to equation (1), given a function f, we define the following indicator function

$$t(x_i \mid x_i, x_j) = \begin{cases} 1 & \text{if } f(x_i) > f(x_j), \\ 0 & \text{if } f(x_i) = f(x_j), \\ -1 & \text{otherwise.} \end{cases} \quad (4)$$

The codomain for t, $\{-1, 0, 1\}$, was chosen for the purpose of the approximation we use later in the section.

In our simplified model we define the information content of a function as a tuple (X, t) including a set X of configurations (the search space) and an indicator function $t : X \times X \to \{1, 0, -1\}$. For every pair (x_i, x_j) of elements in X, t indicates the preference (if any) of the algorithm for one of the solutions. Naturally, the function t can be represented as an $|X| \times |X|$ information matrix M with entries $m_{i,j} = t(x_i, x_j)$.

It is important to note that not all information matrixes can be associated to a fitness function (the information matrix not necessarily represents a partial order). We will call *invalid* those information vectors that cannot be derived from corresponding fitness landscapes.

Since $t(x_i, x_j) = -t(x_j, x_i)$, the matrix M presents symmetries with respect to the main diagonal which reduces the number of available degrees of freedom to the elements above (or below) the diagonal. So, in order to represent an information matrix more concisely, we use the following vector to store the relevant (above diagonal) entries in the information matrix:

$$V = (v_1, v_2, ..., v_n) = (m_{1,2}, m_{1,3}, ..., m_{|X|-1,|X|}), \tag{5}$$

where $n = (|X|^2 - |X|)/2$. Throughout the paper, we use the matrix notation only to illustrate, graphically, some concepts, otherwise we use the vector notation.

3.1 A First Order Approximation

The performance function $\mathbf{P}$, using our model, is a function of V and, as discussed in section 2, the target solution, x_{target}. That is, $\mathbf{P} : V \times X \to \mathbb{R}$. This may be a very complicated function. However, one might wonder whether a first order approximation

$$\mathbf{P}(V, x_{trgt}) \approx c_0 + \sum_{i=1}^{n} c_i v_i. \tag{6}$$

could be sufficient in order to model, to some extent, the performance of a simple GA.[1] We denote the vector $C = \{c_i\}$ as the *performance vector*. The approximation to the performance can be written in a vector notation as:

$$P(V, x_{trgt}) \approx c_0 + C \cdot V \tag{7}$$

In order to calculate the coefficients c_i we can apply statistical or machine learning techniques. For a fixed x_{target}, a training set is made up of pairs of the form $(V_k, \mathbf{P}_k)$, $k = 1, 2, \cdots$, where V_k is a particular information vector – an input for the learner – and $\mathbf{P}_k$ is the corresponding performance measure – the target output for the learner. Ideally, we would want $\mathbf{P}_k = E[\mathbf{P}(V_k)]$ (where the expectation is over multiple runs). Since we do not know the function $\mathbf{P}(V)$, we need to obtain the target values $\mathbf{P}_k$ by some other means. These values, for example, can be estimated by averaging the performance recorded by running the algorithm a suitably large number of times over the particular landscape in question. In order to estimate the coefficients c_i we apply multivariate linear regression over our training set.

Because of the dimensionality of C, this approach can only tackle small landscapes (e.g., 3 loci). In the following section, however, we develop an approach based on a notion of distance which allows us to apply our model for bigger landscapes (14 bits). In principle, an indication to the performance can be obtained also for realistic landscapes. However, in this case, one has to *sample* V and so the sampling noise is compounded with the errors already present due to the linearity of the approximation, resulting in unacceptable errors. As providing a new predictive measure of problem difficulty is not our main objectives, we do not explore scalability issues further in this paper.

---

[1] The purpose of approximating V rather than $\mathcal{P}_f$, is first and foremost, to be able to validate the first order approximation *empirically*.

3.2   A Predictive Measure of Problem Difficulty

In section 2.2 a distance between $\mathcal{P}_f$ and an optimal $\mathcal{P}_{f_{opt}}$ was suggested as an indication to problem difficulty. However, this distance cannot be computed in practice. The model, V, makes it possible to calculate this distance explicitly. In this section we give an indication to problem difficulty which is based on a distance between an information vector, V, and an optimal information vector V_{max}. We conclude this section by arguing that instead of V_{max}, which is hard to compute, we can use the information vector of a known easy problem, making it possible to estimate the distance without calculating the performance vector C.

We begin by assuming that C is known. Therefore it is easy to construct an information vector $V_{\max} = (v_{\max_1}, \cdots, v_{\max_n})$ which contains only positive information. The performance of an algorithm on such a landscape is maximal:

$$v_{\max_i} = \arg\max_{v_i}[c_i v_i], \qquad (8)$$

Optimal information vectors for a given set of coefficients c_i (algorithm) are those where $v_i = 1$ for all i where $c_i > 0$, $v_i = 0$ for all i for which $c_i < 0$, and v_i takes any value for all the remaining i's. The worst possible landscapes, $V_{\min}$ are constructed similarly (note $v_{\min_i} = -v_{\max_i}$ for all i where $c_i \neq 0$).

Given a landscape V, each entry v_i for which $v_i \neq v_{max_i}$, gives an indication to the expected difficulty of V. More generally, *the number of non-matching entries between V and $V_{\max}$ is a rough indicator of problem difficulty.* This is only an indicator because we do not consider the magnitude of the coefficients c_i, only their sign. We define the number of non-matching entries between two landscapes V_1, V_2 as the distance $d(V_1, V_2)$:

$$d(V_1, V_2) = \frac{1}{n} \sum_i |v_{1_i} - v_{2_i}|. \qquad (9)$$

For landscapes without any 0 elements, the distance between two landscapes is the proportion of non-matching entries in the two vectors representing the landscapes.

The distance $d(V_{\max}, V)$ provides an indication to the difficulty of V. However, the set of coefficients C cannot be calculated for realistic problems, and hence $V_{\max}$ cannot be calculated. Instead, in the empirical validation we use an estimation of $V_{\max}$. This can be any landscape which is known to be very easy to optimize, V_{easy} (e.g., ONEMAX for GA). Once V_{easy} is given, the distance can be calculated and the hardness approximated. More formally, we propose to use as an *indicator of problem difficulty* the quantity

$$h(V) = d(V, V_{easy}), \qquad (10)$$

where h is mnemonic for "hardness" and $d(\cdot, \cdot)$ is a distance measure between landscapes.

$h(V)$ gives an indication to problems hardness. The precision of this indicator depends, first of all, on the first order approximation assumption. Moreover, since it does not consider the magnitude of the coefficients c_i, it depends also on their distribution. Finally, since we use V_{easy} to approximate V_{max} it also depends on the distance, $d(V_{max}, V_{easy})$, between the two. Despite these numerous approximations, as we will show in section 4.2 the approach produced very good results, suggesting that all approximations (including the pseudo-linearity of the performance function) are reasonable.

3.3   Indication for the Expected Entropy

In section 2.3 we argued that the number of ties for stochastic search algorithms can be measured as the expected entropy of the distribution P_f^i. However, this cannot be done in practice. Using the vector V we suggest to use as a replacement for the expected entropy the average number of pairwise ties, denoted $H(V)$, which can be easily calculated:

$$H(V) \equiv \frac{1}{n} \sum_i \delta(v_i = 0)$$

In the following we will still refer to $H(V)$ as the entropy of an information vector.

The first order approximation defined in the previous section suggests that the entropy gives an upper bound to the magnitude to which the algorithm can perform either better or worse than a random search. Performance is measured as $P(V) = c_0 + \sum v_i c_i$. $H(V)$ counts the number of entries with a value equal to 0. Clearly, the larger $H(V)$ the smaller the deviation $|P(V) - c_0|$ of the performance from random search can be.

$H(V)$ can be used in practice to calculate an indication to the entropy of the fitness function. This can be done by estimating the fitness distribution of the function, for which several methods exists (e.g., [20]). Once this is done, the number of solutions with equal fitness values can be estimated. We already gave in section 2.3 examples to NP-hard problems with maximum entropy. Intermediate values of entropy can help to tune the exploration–exploitation tradeoff of the algorithm: the higher $H(V)$ the more explorative (or randomized) the algorithm is. Presumably, the mutation rate for such functions should be smaller (and the other way around). We plan to investigate this in future research.

4   Empirical Evaluation

It is hard to assess mathematically the accuracy of our framework. This is because the framework is applicable to search algorithms in general, but, potentially, each algorithm has a different performance function $P(V)$. In addition, it is exceptionally hard to build an explicit formulation for this function, even when considering a specific search algorithm (e.g., a GA). So, empirical validation is the only viable strategy.

Focusing on GAs, in this section, we describe empirical evidence which strongly corroborates the framework. In particular, we show that the performance vector can be used in order to predict the performance of the algorithm on unseen problems. In Section 4.1, we conduct an exhaustive analysis on small landscapes (for the search space of binary strings of length 3) and show that this is indeed the case. Then, moving away from landscapes of a small size towards more realistic sizes (14 bits), in Section 4.2 we use various examples of known problems from the literature (e.g., multi modal landscapes, NIAH, MAXSAT, etc.) in order to show that the hardness of a problem can be estimated by measuring its distance from an easy reference landscape using Equation 10.

4.1   Exhaustive Analysis

In this section we test our main hypothesis and we show that our framework can be used in order to estimate the performance of a GA. Since this requires a full knowledge of the performance vector, we provide results for small landscapes.

We used a simple GA with one-point crossover applied with 100% probability. The takeover time (i.e., the time required for the entire population to converge to the target solution) was used as the performance measure (that is, in this case, we consider minimization). We used a population size of 14. The maximum number of generations was 500. The search on each landscape was repeated 1000 times so as to obtain accurate averages. The target solution (global optimum) was excluded from the first generation.

The experimental setup might look unusual. What is the purpose of using a population of size 14 and 500 generations to explore a search space of size 8? We decided to choose these settings in order to get higher resolution for the performance of a GA. For this purpose, the performance measure we chose is the takeover time (rather than, for example, the number of generations it takes to sample the optimum). The takeover time depends, once the optimum is sampled, on the selection pressure. However, firstly, we assume that the easier the landscape, the more copies of the optimum will be in early generations and so the faster the takeover time will be and secondly, since we use a very low selection pressure (tournament of size 2), the influence of a random occurrence of an optimum will not be crucial.

In a first experiment, we measured the mean takeover time for *all possible valid landscapes*. These are landscapes which can be derived from a fitness function and where none of the elements is 0. It is important to emphasize that the target solution was fixed. Therefore, we were measuring the performance of a GA on all possible landscapes given that the optimum is at a particular position in the search space. Since the size of the search space X is 8, the reduced search space X' is of size 7 and, so, we have $7! = 5040$ possible landscapes. In order to estimate the performance vector (Equation 6) we used multivariate linear regression on the results obtained from running the GA over all such landscapes. The correlation coefficient between observed and predicted performance is 0.935.

In order to verify whether the linear approximation to $P(V)$ generalizes well, we sampled 1000 additional landscapes out of the entire space of possible information vectors (i.e., including invalid landscapes, see Section 3). The correlation between prediction and observation is still very high (0.923), suggesting good generalization. For more details about the multivariate linear regression the co-efficients obtained and a possible interpretation see [3].

4.2 Estimation of Problem Hardness

In the previous section we have shown that our framework can be used in order to accurately estimate the performance of a GA and assess problem difficulty. However, it is clear that a direct estimation of the performance vector can only be used for small search spaces. In this section, rather than using the performance vector directly, we use the ideas presented in Section 3.2 to see if the approach can be applied to more practical scenarios.

In Section 3.2 we argued that the hardness of a problem can be estimated using the distance of its information vector from the optimal landscape $V_{\max}$. Since, in general, the optimal landscape is not known, we proposed to use an approximation, V_{easy}, instead (Equation 6). The question now is which easy problem to choose. We know from many empirical studies that unimodal problems tend to be GA-easy, the Onemax problem being a glowing example. The Onemax belongs to the following more general class of functions: $f(x) = \sum_i \delta(x_i = x_{\text{target}_i})$ where x_{target} is the global optimum. Onemax, is a specific case, where x_{target} is the string of all ones. In this work we decided to use the information vector V_{easy} derived from $f(x)$ as an approximation of the optimal landscape $V_{\max}$.

The information vector and the performance function are defined for a fixed target solution. If we change the target solution the same information vector can change from being easy to being difficult (e.g., consider the information vector induced by the Onemax function where we change the optimum to being the string 000). So, the distance between landscapes must be computed for landscapes with the same global optimum. This requires knowing the global optimum in advance.

In the following experiments, we calculated $h(V)$ as the distance between the actual landscape induced by a problem and the one induced by the function $f(x)$ using the global optimum of the problem as x_{target}. We used a simple GA with uniform crossover applied with 100% probability and mutation applied with 10% probability. The search space included binary strings of 14 bits. The population size is 20. The first generation in which the optimum was found was used as the performance measure. The results are averages of 100 runs.

The remainder of this section is organised as follows. First we give empirical results for various problems then we test our approach on three counterexamples for other measures of problem difficulty.

Hardness of standard test problems. In this subsection we estimate the hardness of problems with no information (NIAH), random information or random problems (RAND), maximally reliable information (unimodal functions) and maximally unreliable information (deceptive functions). Furthermore, we study problems with a variable level of difficulty: the NK landscapes with $k =1$–10, multimodal landscapes (MM1, MM2, etc.) with a varying number of local maxima (1–20), and trap functions ($TRAP_i$ where $i \in \{1, \cdots, l\}$ indicates the level of difficulty, $TRAP_1$ being the most difficult). Finally, to test our measure of difficulty on landscapes which were not induced by artificial problems, we also considered 12 random MAXSAT problems (14 literals, 59 clauses). For each problem, we consider only one global optimum. If a problem had more than one global optimum, we chose one at random to be the target solution.

Table 1 gives our (predicted) measure of difficulty $(h(V))$ for selected problems and the actual performance obtained by the GA (P). Note that $h(V)$ is scaled between 0 (very easy) and 1 (very hard) while the performance P between 1 (when, on average, the optimum was sampled in the first generation) and 100 (on average, the optimum was not sampled in 100 generations).

As one can observe, the scale of $h(v)$ is neither linear nor always consistent. For example, a difference of only 0.002 between MM5 (0.403) and NK2 (0.405) corresponds to a very large difference in performance: 68.7 for MM5 vs. 33.9 for NK2. Still, the correlation coefficient between observed and predicted difficulty (for all problems) is 0.82.

The table confirms some of the previous results regarding GA hardness (see section 1). For example, the table shows that multimodality is not a good indicator of problem difficulty [11]. A landscape with 9 local maxima has the same expected difficulty as a landscape with 16 local maxima, while a landscape with 5 local maxima is more difficult than a landscape with 16.

Table 1. Estimated hardness $h(V)$ and average performance P for a selection of 40 test problems

$h(V)$	P	Problem	$h(V)$	P	Problem	$h(V)$	P	Problem	$h(V)$	P	Problem
0.000	11.4	MM1	0.363	77.3	MM3	0.405	71.4	MM12	0.493	87.9	RAND
0.226	40.1	MM2	0.372	52.8	MM6	0.422	90.5	NIAH	0.496	83.6	NK9
0.293	23.0	NK1	0.374	59.3	MAXSAT	0.427	52.9	NK4	0.497	88.0	NK6
0.306	27.5	MAXSAT	0.374	58.8	MM9	0.44	74.8	MM13	0.502	89.2	RAND
0.325	28.2	MAXSAT	0.384	60.4	MM16	0.452	70.1	MM7	0.507	90.7	MM14
0.325	48.7	MAXSAT	0.387	53.7	MAXSAT	0.463	79.5	MM17	0.509	89.0	NK7
0.334	56.9	MM4	0.393	55.7	MM11	0.463	65.5	ALTNBR	0.511	87.0	NK8
0.336	39.3	MAXSAT	0.394	51.1	MAXSAT	0.473	82.9	MM15	0.605	79.3	TRAP4
0.355	42.6	MAXSAT	0.403	68.7	MM5	0.473	76.01	NK3	0.756	99.0	TRAP3
0.361	51.4	MAXSAT	0.405	33.9	NK2	0.491	90.4	MM18	0.842	100	TRAP1

The table also confirms that the NK model is not appropriate for the study of problem difficulty because problems with a $k > 2$ are already very difficult [11]. Indeed, our measure suggests that the difficulty of such landscapes is close to random.

Different instances of the same problem might have different degrees of difficulty in the black-box scenario [11]. Indeed, the predicted difficulty for different instances of the MAXSAT problems varies from 0.306 (easy) to 0.394 (difficult) – even though, all were chosen with the same variable to clause ratio.

Hardness of known counterexamples for other difficulty measures. In the previous section we presented evidence supporting the hypothesis that $h(V)$ is a meaningful indicator of problem difficulty. In this section we test our framework on three problems where other measures of difficulty have been shown to fail.

Naudts and Kallel [16] constructed a simple problem consisting of a deceptive mixture of Onemax and Zeromax, where both the fitness distance correlation (FDC) measure and the sitewise optimisation measure (a generalisation for the FDC and epistasis suggested in the same paper) failed to correctly predict performance. For this class of problems, the higher the mixture coefficient m, the harder the problem, yet no hardness measures was able to predict this. We performed experiments with this problem[2], with the control parameter m varying from 1 (easy) to 9 (hard). The correlation between the predicted difficulty and the actual performance was 0.75. So, we were largely able to capture the difference in performance as the parameter m varied.

Jansen [11] showed that the fitness distance correlation of the ridge function is very small. Yet, this is an easy problem for a hill climber. So, we decided to apply our method to this function as well. The distance of the ridge function to the optimal landscape is 0.84, which indicates a very difficult problem. Indeed, the GA was not able to find the solution in 100 generations. A problem that is easy for a hill climber is not necessarily easy for a recombinative GA.

Jansen [11] gave two counter examples to the bit-wise epistasis measure of difficulty. The first one was the NIAH which we already discussed before. The second was the leading-one function. The distance of the leading one function from the optimum landscape is 0.36, which predicts well the performance shown by a GA (50.3 average number of generations required to sample the global optimum).

5   Conclusions

The decomposition of f into $\mathcal{P}_f$ captures the way in which the algorithm uses the fitness function – i.e., via the selection paradigm – to define a probability distribution over sampled points which is used to pick the solutions around which to expand the search.

This decomposition enables one, to a large extent, to approximate the performance of a GA using a simple, linear function. In section 4.1 we demonstrated this empirically for problems of small size (3 bits). Also, a predictive measure based on the linear approximation was used to assess the GA hardness (section 4.2) of several larger problems (14 bits).

Ties in classical graph search algorithms are known to lead to poor performance. We assessed the influence of ties for RSHs using the expected entropy of $\mathcal{P}_f$. Interestingly, the linear approximation predicts that the number of ties as measured in section 3.3 gives a bound to the expected performance. In the extreme case where the entropy is maximal performance cannot be better than that of random search.

---

[2] The GA used in [16] is different from the one used here.

While the framework presented in section 2 is general, the first order approximation was tested only for GAs. We expect the same approximation to hold for other population based, stochastic algorithms. However, this remains to be checked.

The drawbacks of this framework is the lack of bounds to its precisions. The approximation suggested in this paper and the conclusions which can be derived from it can be used only as a first indication to the properties of particular algorithms or functions. Nevertheless, the ever increasing number of new complex metaheuristics makes a rough, but quick estimation a necessity.

Acknowledgements

The authors would like to thank the anonymous reviewers for their valuable suggestions, which have improved the paper greatly.

References

1. Altenberg, L.: Evolving better representations through selective genome growth. In: Proceedings of the 1st IEEE Conference on Evolutionary Computation, Orlando, Florida, USA, June 27-29, 1994, vol. 1, pp. 182–187. IEEE, New York (1994)
2. Altenberg, L.: NK fitness landscapes. In: Handbook of Evolutionary Computation, pp. B2.7.2. Oxford University Press, Oxford (1997)
3. Borenstein, Y., Poli, R.: Information landscapes and the analysis of search algorithms. In: GECCO '05. Proceedings of the 2005 conference on Genetic and evolutionary computation, New York, NY, USA, pp. 1287–1294. ACM Press, New York (2005)
4. Davidor, Y.: Epistasis variance: A viewpoint on GA-hardness. In: Rawlins, G.J.E. (ed.) Proceedings of the First Workshop on Foundations of Genetic Algorithms, Bloomington Campus, Indiana, USA, July 15-18, 1990, pp. 23–35. Morgan Kaufmann, San Francisco (1990)
5. Droste, S., Jansen, T., Wegener, I.: Upper and lower bounds for randomized search heuristics in black-box optimization. Electronic Colloquium on Computational Complexity (ECCC) (048) (2003)
6. Forrest, S., Mitchell, M.: Relative building-block fitness and the building block hypothesis. In: Whitley, L.D. (ed.) Proceedings of the Second Workshop on Foundations of Genetic Algorithms, Vail, Colorado, USA, July 26-29, 1992, pp. 109–126. Morgan Kaufmann, San Francisco (1992)
7. Rothlauf, F.: Representations for Genetic and Evolutionary Algorithms. Springer, Heidelberg (2002)
8. Goldberg, D.E.: Genetic Algorithms in Search, Optimization, and Machine Learning. Addison-Wesley, London (1989)
9. Goldberg, D.E.: Making genetic algorithm fly: a lesson from the wright brothers. Advanced Technology For Developers 2, 1–8 (1993)
10. Grefenstette, J.J.: Deception considered harmful. In: Whitley, L.D. (ed.) Proceedings of the Second Workshop on Foundations of Genetic Algorithms, Vail, Colorado, USA, July 26-29, 1992, pp. 75–91. Morgan Kaufmann, San Francisco (1992)
11. Jansen, T.: On classifications of fitness functions. In: Theoretical aspects of evolutionary computing, pp. 371–385. Springer, London, UK (2001)

12. Jones, T., Forrest, S.: Fitness distance correlation as a measure of problem difficulty for genetic algorithms. In: Proceedings of the 6th International Conference on Genetic Algorithms, San Francisco, CA, USA, 1995, pp. 184–192. Morgan Kaufmann Publishers Inc. San Francisco (1995)

13. Kauffman, S.: The Origins of Order: Self-Organization and Selection in Evolution. Oxford University Press, Oxford (1993)

14. Moraglio, A., Poli, R.: Topological interpretation of crossover. In: Deb, K., et al. (eds.) GECCO 2004. LNCS, vol. 3102, pp. 1377–1388. Springer, Heidelberg (2004)

15. Naudts, B.: Measuring GA-hardness. PhD thesis, University of Antwerpen, Antwerpen, Netherlands (1998)

16. Naudts, B., Kallel, L.: A comparison of predictive measures of problem difficulty in evolutionary algorithms. IEEE Trans. Evolutionary Computation 4(1), 1–15 (2000)

17. Pearl, J.: Heuristics: intelligent search strategies for computer problem solving. Addison-Wesley Longman Publishing Co., Inc. Boston, MA, USA (1984)

18. Radcliffe, N.J.: Equivalence class analysis of genetic algorithms. Complex Systems 5, 183–205 (1991)

19. Rana, S.: Examining the Role of Local Optima and Schema Processing in Genetic Search. PhD thesis, Colorado State University, Colorado, U.S.A (1998)

20. Rose, H., Ebeling, W., Asselmeyer, T.: The density of states - a measure of the difficulty of optimisation problems. In: Parallel Problem Solving from Nature, pp. 208–217 (1996)

21. Vose, M.D.: The Simple Genetic Algorithm: Foundations and Theory. MIT Press, Cambridge, MA, USA (1998)

On the Effects of Bit-Wise Neutrality on Fitness Distance Correlation, Phenotypic Mutation Rates and Problem Hardness

Riccardo Poli and Edgar Galván-López

Department of Computer Science
University of Essex
Colchester, CO4 3SQ, UK
{rpoli,egalva}@essex.ac.uk

Abstract. The effects of neutrality on evolutionary search are not fully understood. In this paper we make an effort to shed some light on how and why bit-wise neutrality – an important form of neutrality induced by a genotype-phenotype map where each phenotypic bit is obtained by transforming a group of genotypic bits via an encoding function – influences the behaviour of a mutation-based GA on functions of unitation. To do so we study how the fitness distance correlation (*fdc*) of landscapes changes under the effect of different (neutral) encodings. We also study how phenotypic mutation rates change as a function of the genotypic mutation rate for different encodings. This allows us to formulate simple explanations for why the behaviour of a GA changes so radically with different types of neutrality and mutation rates. Finally, we corroborate these conjectures with extensive empirical experimentation.

1   Introduction

Evolutionary Computation (EC) systems are inspired by the theory of natural evolution. The theory argues that through the process of selection, organisms become adapted to their environments and this is the result of accumulative beneficial mutations. However, in the late 1960s, Kimura [22] put forward the theory that the majority of evolutionary changes at molecular level are the result of random fixation of selectively neutral mutations. In other words, the mutations that take place in the evolutionary process are neither advantageous nor disadvantageous to the survival of individuals. Kimura's theory, called neutral theory of molecular evolution, considers a mutation from one gene to another as neutral if this modification does not affect the phenotype.

Within the context of EC, different approaches have been proposed to study neutrality in evolutionary search. Whether or not neutrality helps evolutionary search, however, has not conclusively been established. In the following section, we will present work that shows clearly that the relationship between the genotype space and phenotype space when neutrality is present in the evolutionary search plays a crucial role.

C.R. Stephens et al. (Eds.): FOGA 2007, LNCS 4436, pp. 138–164, 2007.

The aims of our work are:

- understanding the relationship between the solution space (represented at phenotype level) and the search space (represented at genotype level) in the presence of neutrality, and, following this analysis,
- identifying under what circumstances neutrality may help to improve performance of evolutionary processes.

The paper is organised as follows. In the next section, previous work on neutrality in EC is summarised. In Section 3, we describe the genotype-phenotype encodings studied in this paper. In Section 4 we review the notion of the fitness distance correlation (*fdc*), introduce our test problems and look at their *fdc* in the absence of neutrality. In Section 5 we study the effects of bitwise neutrality on the difficulty of our test problems, exploring the case of the OneMax problem in particular depth. Section 6 makes the relation between genotypic mutation rates and phenotypic mutation rates explicit. In Sections 7 and 8 we present and discuss the results of experiments with unimodal, multimodal and deceptive landscape problems and draw some conclusions.

2   Previous Work

In biology the effects of neutrality have been extensively discussed in numerous studies (see for example [22,18,13]). As a consequence of these studies, there has been a growing interest in using and analysing the effects of neutrality in EC.

For instance, Barnett [3] introduced a new family of *NK* fitness landscapes that has the property of allowing the explicit addition of neutrality in the evolutionary process. He called them *NKp* fitness landscapes. The parameter p is determines the amount of neutrality that is be present during evolution ($p = 0$ corresponds to a normal *NK* landscape while $p = 1$ corresponds to a flat landscape). Barnetts' motivation was to see if the constant innovation property observed in biology [18] was present in this type of neutrality. The author argued that, at least for a mutation-selection algorithm, avoiding to get stuck in local optima can be achieved in the presence of neutral networks.

In a insightful investigation, Weicker and Weicker [36] used two methods to analyse the effects of redundancy: diploid and decoders. For the former method, each individual contains two solutions and an extra bit which is in charge to set the active solution. So, it is clear that the size of the search space for this kind of redundancy has increased dramatically. Moreover, this kind of mapping is homogenous. This is not the case, however, for the decoder method. A decoder is effectively a repair mechanism that maps an invalid genotype (e.g., one that violates some constraints) into a valid one. They investigated the effects of both methods with respect to local optima, finding that local optima in the search space are converted to plateaus. However, this does not mean that this represent an advantage, as the authors pointed out saying: "... redundancy has many facets with various different characteristics. The mapping from those characteristics to the expected performance remains to be done" [36].

Toussaint and Igel [33] pointed out that standard approaches to self-adaptation [10] are a basic and explicit example for the benefit of neutrality. In these approaches the genome is augmented with strategy parameters which typically parameterise the mutation distribution (e.g., the mutation rate). These neutral parts of the genome are co-adapted during evolution to induce better search distributions. Interestingly, theoretical work on the evolution of strategy parameters [4] can so be re-interpreted as theoretical results on the evolution of neutral traits. The point of view developed in [33] conversely suggests that the core aspect of neutrality is that different genomes in a neutral set provide a variety of different mutation distributions from which evolution may select in a self-adaptive way.

This line of thought was further formalised by Toussaint [32]. Given a fixed GP-map one can investigate the varieties of mutation distributions induced by different genomes in a neutral set. If their phenotypic projections (the phenotypic mutation distributions) are constant over each neutral set this is defined as *trivial neutrality*. Toussaint shows that trivial neutrality is a necessary and sufficient condition for compatibility with phenotypic projection of a mutation-selection GA. Intuitively this means that, in the case of trivial neutrality, neutral traits have no effect on phenotypic evolution. I.e., whether one or another representative of a neutral set is present in a population does not influence the evolution of phenotypes. Note that one of the encodings we will investigate (the Parity encoding) is a case of trivial neutrality. This and calculations presented in Section 5 will help us explain the results presented in Section 7. In the case of non-trivial neutrality, different genotypes in a neutral set induce different phenotypic distributions, which implies a selection between equivalent genotypes similarly to the selection of strategy parameters in self-adaptive EAs. Toussaint interprets this as the underlying mechanism of the evolution of genetic representations.

Vassilev and Miller [35] claimed that the presence of neutrality in evolutionary search was useful when they used Cartesian Genetic Programming (CGP) [25] to evolve digital circuits. For their study, the authors considered the well-known three-bit multiplier problem. They focused their attention on the relation between the size and the height of the landscapes plateaus. In their work, Vassilev and Miller suggested that the length of neutral walks will decrease as the best fitness increases. They concluded that neutrality helps to cross wide landscapes areas of low fitness.

Smith *et al.* [29] analysed the effects of the presence of neutral networks on the evolutionary process. They observed how evolvability was affected by the presence of such neutral networks. For their study they used a system with an extremely complex genotype-to-fitness mapping. They concluded that the existence of neutral networks in the search space, which allows the evolutionary process to escape from local optima, does not necessarily provide any advantage. This is because the population does not evolve any faster due to inherent neutrality. In a different piece of work [30], the same authors focused their research on looking at the dynamics of the population rather than looking at just the fitness, and argued that neutrality did not perform a useful role in an evolutionary robotic task.

Ebner *et al.* [9] studied the effects of neutrality on the search space. For this purpose, they separate the search space into phenotypes which belong to different species. In their work, they proposed three different types of encodings which, according to the authors, seem to allow a high degree of connectivity among neutral networks and so, individuals will not have problem discovering other species. From their experiments, they concluded that the higher the degree of redundancy (another term for neutrality) is, the better species are able to adapt. In other words, redundancy avoids getting stuck in local optima.

Yu and Miller [37] showed in their work that neutrality improves the evolutionary search process for a Boolean benchmark problem. They used Miller's CGP [25] to measure explicit neutrality in the evolutionary process. They explained that mutation on a genotype that has part of its genes active and others inactive may produce different effects: mutation on active genes is adaptive because it exploits accumulated beneficial mutations, while mutation on inactive genes has a neutral effect on a genotype's fitness, yet it provides exploratory power by maintaining genetic diversity. Yu and Miller extended this work in [38] showing that neutrality was helpful and that there is a relationship between neutral mutations and success rate in a Boolean function induction problem. However, Collins [7] claimed that the conclusion that neutrality is beneficial in this problem is flawed.

Yu and Miller also investigated neutrality using the simple OneMax problem [39]. They attempted a theoretical approach in this work. With their experiments, they showed that neutrality is advantageous because it provides a buffer to absorb destructive mutations.

Chow [5] studied the relationship between genotype space and phenotype space. In his work, Chow used a hybrid algorithm (a GA receiving feedback from a hill climber). The approach proposed by Chow relies on replacing a genotype by converting a phenotype to its corresponding genotype. Such phenotype is given to the GA by the hill climber. Chow claimed that in all experiments, such replacements improved the search results.

Fonseca and Correia [12] developed a mathematical model which is able to include the properties proposed by Rothlauf and Goldberg [28] and which are claimed to influence the quality of redundant representations. All their experiments were carried out in the context of a simple mutation-selection evolutionary model. Under this model, the authors were wondering whether a redundant representation might be constructed which preserves evolutionary behaviour. Based on their mathematical model, they claimed that the presence of non-coding genes do not affect the evolutionary process. However, they were unable to determine what kind of representation (redundancy) is necessary to obtain good results on a given optimisation problem.

Banzhaf and Leier [2] studied the effects of neutral networks' connectivity using a Boolean problem. They studied the effects of neutrality using 2 NAND space and showed how it can aid the evolutionary search. For this purpose,

Banzhaf and Leier used a linear GP representation because, they argued, with GP it is easier to identify neutrality than in other evolutionary method. In their experiments and by means of an exhaustive examination of all possible genotypes, they showed how there are highly common phenotypes and very few uncommon phenotypes. The authors concluded that neutral networks must be highly intertwined to allow a quick transition from one network to the next.

In summary, the literature presents a mixed picture as to what the effects of neutrality on evolutionary search are.

As can be seen from previous paragraphs, the relationship between phenotype and genotype space is crucial to understand the influence of neutrality on evolutionary search. We believe that the effects of neutrality on evolutionary search are not well understood for several reasons:

- studies often consider problems, representations and search algorithms that are relatively complex and so results represent the compositions of multiple effects (e.g., bloat or spurious attractors in genetic programming),
- there is not a single definition of neutrality and different studies have added neutrality to problems in radically different ways, and,
- the features of a problem's landscape change when neutrality is artificially added, but rarely an effort has been made to understand in exactly what ways.

Recently [14,24], in an effort to shed some light on neutrality we started addressing these problems. In particular, we studied perhaps the simplest possible form of neutrality: a neutral network of constant fitness, identically distributed in the whole search space. For this form of neutrality, we analysed both problem-solving performance and population flows from and to the neutral network and the basins of attraction of the optima, as the fitness of the neutral network was varied.

In this paper, we will continue towards the same goals, but we will consider a much more practical form of neutrality, bit-wise neutrality.

3   Bitwise Neutrality

Bitwise neutrality is a form of neutrality induced by a genotype-phenotype map where each phenotypic bit is obtained by transforming a group of genotypic bits via some encoding function. In this paper we consider three different kinds of genotype-phenotype encodings to specify bitwise neutrality in the evolutionary process. For the three of them, each phenotypic bit is encoded using n genotypic bits.

These encodings are defined as follows:

1. The *majority* encoding works as follows: given n bits, the user defines a threshold (T) $(0 \leq T \leq n)$ and if the number of ones that are in the n genotypic bits is greater or equal to T then the bit at the phenotype level is set to 1, otherwise it is set to 0. Figure 1(a) illustrates this concept. Normally,

to avoid biasing the system we will always use $T = n/2$ and n odd. This guarantees that 0s and 1s are treated identically.

2. The *parity* encoding works as follows: if the number of ones that are in n genotypic bits is an even number, then the bit at the phenotype level is set to 1, otherwise it is set to 0. Figure 1(b) illustrates this concept.
3. The *truth table* encoding works as follows: a truth table is generated and the output for each combination is produced at random. (Half of the outputs of the truth table are assigned with 0s and the other half are assigned with 1s. Then the outputs are shuffled to make them perfectly random). Then we consider the n genotypic bits as inputs, and we take as our phenotypic bit the corresponding truth table's output. Figure 1(c) illustrates this concept.

Neutrality is added to the non-redundant code by the proposed encodings. Because each bit is encoded using n bits, the same phenotype can be obtained from different genotypes and, so, neutrality is artificially added to the search space.

In the presence of the form of neutrality discussed above, the size of the search space is $2^{\ell n}$, where ℓ is the length of a phenotypic bit string and n is the number of bits required to encode each bit. With the types of encodings explained earlier, we have increased not only the size of the search space but also the size of the solution space. However, this does not mean that neutrality is always beneficial. We have also to bear in mind that the mutation rate at genotype level is different than the mutation rate at phenotype level. We will calculate these mutations rates and see their effects in Section 6.

Neutrality is often reported to help in multimodal landscapes, in that it can prevent a searcher from getting stuck in local optima. However, very little mathematical evidence to support this claim has been provided in the literature. So, in the next section we start our analysis by using a well-defined hardness measure, the fitness distance correlation, calculating it in such a way to make the dependency between problem difficulty and neutrality of the encoding explicit.

4   Fitness Distance Correlation

4.1   Definition

Jones [19,20] proposed *fitness distance correlation* (fdc) to measure the difficulty of problem by studying the relationship between fitness and distance. The idea behind *fdc* was to consider fitness functions as heuristics functions and to interpret their results as indicators of the distance to the nearest global optimum of the search space and, so, *fdc* is an algebraic measure to express such a relationship.

The definition of *fdc* is quite simple: given a set $F = \{f_1, f_2, ..., f_n\}$ of fitness values of n individuals and the corresponding set $D = \{d_1, d_2, ..., d_n\}$ of distances to the nearest global optimum, we compute the correlation coefficient r, as:

$$r = \frac{C_{FD}}{\sigma_F \sigma_D},$$

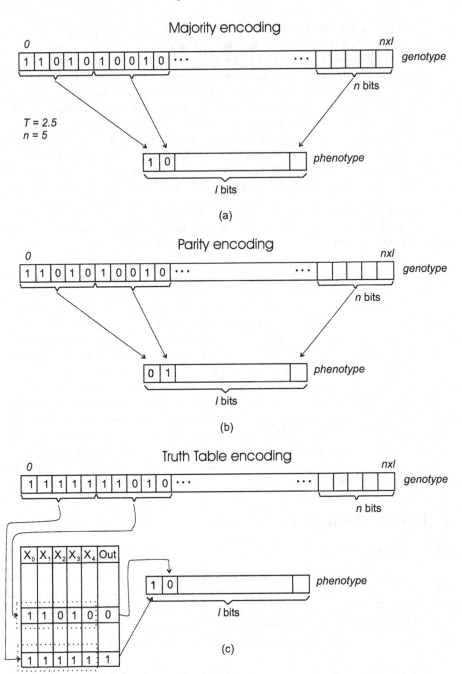

Fig. 1. Three different encodings used in our research: (a) Majority encoding, (b) Parity encoding and (c) Truth table encoding

where:

$$C_{FD} = \frac{1}{n} \sum_{i=1}^{n} (f_i - \overline{f})(d_i - \overline{d})$$

is the covariance of F and D, and σ_F, σ_D, $\overline{f}$ and $\overline{d}$ are the standard deviations and means of F and D, respectively. The n individuals used to compute fdc can be chosen in different ways. For reasonably small search spaces or in theoretical calculations it is often possible to sample the whole search space. In this case fdc can be computed exactly. However, in most other cases, fdc is estimated by constructing the sets F and D via some form of random sampling.

According to [20] a problem can be classified in one of three classes, depending of the value of r: (1) *misleading* ($r \geq 0.15$), in which fitness tends to increase with the distance from the global optimum, (2) *difficult* ($-0.15 < r < 0.15$), for which there is no correlation between fitness and distance, and (3) *easy* ($r \leq -0.15$), in which fitness increases as the global optimum approaches.

There are some known weakness in the fdc as a measure of problem hardness [1,26,27]. However, it is fair to say that the method has been generally very successful [6,20,31,34]. The distance used in the calculations is, for binary search spaces, the Hamming distance, H.

In this work we will use fdc to evaluate problem difficulty with and without neutrality. Here we only consider problems where the fitness function is a function of unitation, so, we can rewrite C_{FD} in a more explicit form.

4.2 Test Problems

We have used three problems to analyse neutrality. The first one is the OneMax problem. The problem is to maximise:

$$f(x) = \sum_{i} x_i,$$

where x is a binary string of length ℓ, i.e., $x \in \{0,1\}^\ell$. Naturally, this problem has only one global optimum in $11 \cdots 1$, and, the landscape is unimodal. Seen as a function of unitation the problem is represented by $f(u) = u$ or $f(x) = u(x)$ where $u(x)$ is a function that returns the unitation value of x.

For the second problem, we used a multimodal problem generator [8,21,23]. The idea is to create problem instances with a certain degree of multi-modality. In general, for a problem with P peaks, P bit strings of length ℓ are randomly generated. The generator works as follows. To evaluate an arbitrary individual x, we first locate the nearest peak in Hamming space

$$Peak_n(x) = \arg \min_i H(Peak_i, x)$$

In case there is a tie, the highest peak is chosen. The fitness of x is the number of bits the string has in common with that nearest peak, divided by ℓ and scaled by the height of the nearest peak:

$$f(x) = \frac{\ell - H(x, Peak_n(x))}{\ell} \times Height(Peak_n(x))$$

In this problem, the fitness value has a range from 0.0 and 1.0. The goal is to find the highest peak (i.e., to find a string with fitness 1.0). The difficulty of the problem depends on the number of peaks, the distribution of peaks and, finally, the distribution of peak heights. To carry out our experiments, we have tuned the parameters in such a way to make the problem much harder than the OneMax problem but easier than the trap function (see below). More details will be provided in Section 7.

The third and last problem is a Trap function, which is a deceptive function of unitation [15,16,17]. For this example, we have used the function:

$$f(x) = \begin{cases} \frac{a}{u_{min}}(u_{min} - u(x)) & \text{if } u(x) \leq u_{min}, \\ \frac{b}{\ell-u_{min}}(u(x) - u_{min}) & \text{otherwise} \end{cases}$$

where a is the deceptive optimum, b is the global optimum, and u_{min} is the slope-change location. Basically the idea is that there are two optima, a and b, and by varying the parameters ℓ and u_{min}, we can make the problem easier or harder.

4.3   *fdc* in the Absence of Neutrality

For all our test problems, given a search space of binary strings of length ℓ and being the unitation of the optimal string $u_{opt} = \ell$, if we sample the whole search space in order to compute C_{FD}, we have:

$$C_{FD} = \frac{1}{2^\ell} \sum_{u=0}^{\ell} \binom{\ell}{u}(f(u) - \overline{f})(\ell - u - \overline{d})$$

where:

$$\overline{f} = \frac{1}{2^\ell} \sum_{u=0}^{\ell} \binom{\ell}{u} f(u)$$

and

$$\overline{d} = \ell - \frac{1}{2^\ell} \sum_{u=0}^{\ell} \binom{\ell}{u} u = \frac{\ell}{2}.$$

Similar expressions can be obtained for σ_D and σ_F.

So, for example, for OneMax, where $f(u) = u$, we have $\overline{f} = \frac{\ell}{2}$ and

$$C_{FD} = \frac{1}{2^\ell} \sum_{u=0}^{\ell} \binom{\ell}{u}\left(u - \frac{\ell}{2}\right)\left(\frac{\ell}{2} - u\right)$$

$$= -\frac{\ell}{4}$$

as one can easily see by noting that $\frac{1}{2^\ell}\binom{\ell}{u}$ is a binomial distribution, $\binom{\ell}{u}p^u(1 - p)^{\ell-u}$, with success probability $p = 1/2$. Therefore, by definition of variance, $C_{FD} = -Var[u] = -\ell p(1 - p) = -\frac{\ell}{4}$. By similar arguments one finds $\sigma_D^2 = \sigma_F^2 = \frac{1}{4}$, whereby $r = -1$, suggesting an easy problem. For Trap functions, instead, whenever $u_{min} \approx \ell$ one finds $r \approx 1$ [19] indicating hard problems.

5   *fdc* in the Presence of Bitwise Neutrality

As mentioned in Section 3, the form of neutrality we consider here is one where each phenotypic bit is encoded using n genotypic bits. In this situation, C_{FD} is given by:

$$C_{FD} = \frac{1}{2^{n\ell}} \sum_{x \in \{0,1\}^{n\ell}} (f_x(x) - \bar{f})(d(x) - \bar{d})$$

where $x = x_1 \cdots x_{n\ell}$ is a genotype and $f_x(x)$ is the genotypic fitness. Similar expressions can be obtained for σ_D and σ_F. Note that $f_x(x)$ can be written as

$$f_x(x) = f_y(g(x^{(1)}), g(x^{(2)}), \cdots, g(x^{(n)}))$$

where $x^{(k)} = x_{(k-1)n+1} \cdots x_{kn}$ is a substring of x, g is one of our encoding functions (e.g., Majority or Parity), and $f_y(y)$ is the phenotypic fitness ($y \in \{0,1\}^\ell$), which in this work we will assume to be a function of unitation.

We define $X_n = \{x \in \{0,1\}^n : g(x) = 1\}$ and $\bar{X}_n = \{x \in \{0,1\}^n : g(x) = 0\}$. We require that our encoding functions g respect one property: that on average they return as many 0s as 1s, i.e.,

$$\sum_{x \in \{0,1\}^n} g(x) = 2^{n-1}.$$

This property is respected by the encodings described in Section 3. So, $|X_n| = |\bar{X}_n| = 2^{n-1}$.

To illustrate the effects of the introduction of bitwise neutrality, in the following we will consider in detail the case of OneMax.

5.1   *fdc* for OneMax with Bitwise Neutrality

For the OneMax function we have

$$f_x(x) = \sum_i g(x^{(i)}).$$

To compute *fdc* we can make use of a result originally derived by Jones [19, Appendix D]: the concatenation of multiple copies of a problem does not change the *fdc* of the original problem, provided the fitness of the concatenated problem is obtained by summing the fitnesses of the sub-problems. This result is applicable because we can interpret g as the fitness functions of an n-bit problem which is concatenated ℓ times to form an $\ell \times n$ bit problem with fitness function $f_x(x)$. Therefore, we can compute the *fdc* for OneMax with different forms of bitwise neutrality by simply computing the *fdc* of the corresponding g functions. Since these functions take only binary values, this calculation is simpler than the original.

Let us start by considering the mean value of the function g, $\bar{g}$, for all encodings. By definition we have that $g(x) = 1$ for $x \in X_n$ and $g(x) = 0$ otherwise. So, irrespective of the encoding used we have

$$\bar{g} = \frac{1}{2^n} \sum_{x \in \{0,1\}^n} g(x) = \frac{1}{2^n} \sum_{x \in X_n} 1 = \frac{1}{2^n}|X_n| = \frac{1}{2}$$

irrespective of the encoding function used.

We use this result in the computation of σ_F^2, obtaining

$$\sigma_F^2 = \frac{1}{2^n} \sum_{x \in \{0,1\}^n} (g(x) - \bar{g})^2$$

$$= \frac{1}{2^n} \left(\sum_{x \in X_n} \left(1 - \frac{1}{2}\right)^2 + \sum_{x \in \bar{X}_n} \left(0 - \frac{1}{2}\right)^2 \right)$$

$$= \frac{1}{4},$$

which, again, is valid for all encodings.

Also, we have

$$\bar{d} = \frac{1}{2^n} \sum_{x \in \{0,1\}^n} H(x, N(x))$$

where $N(x)$ is the global optimum nearest to x and H is the Hamming distance. Because all elements of X_n are global optima of g, and, so, $x = N(x)$ and $H(x, N(x)) = 0$ for $x \in X_n$, we have

$$\bar{d} = \frac{1}{2^n} \sum_{x \in \bar{X}_n} H(x, N(x)). \tag{1}$$

If we extend the definition of Hamming distance to sets by via the definition $H(x, S) = \min_{y \in S} H(x, y)$, we can rewrite Equation (1) as

$$\bar{d} = \frac{1}{2} E[H(x, X_n)|x \in \bar{X}_n], \tag{2}$$

where $E[H(x, X_n)|x \in \bar{X}_n]$ is the mean Hamming distance between the elements of $\bar{X}_n$ and the set X_n.

Similarly we have,

$$\sigma_D^2 = \frac{1}{2^n} \sum_{x \in \{0,1\}^n} (H(x, N(x)) - \bar{d})^2$$

$$= \frac{1}{2^n} \left(\sum_{x \in X_n} (0 - \bar{d})^2 + \sum_{x \in \bar{X}_n} (H(x, N(x)) - \bar{d})^2 \right)$$

$$= \frac{1}{2} \left(\bar{d}^2 + E\left[\left(H(x, X_n) - \bar{d} \right)^2 \Big| x \in \bar{X}_n \right] \right)$$

$$= \frac{1}{2} \left(\bar{d}^2 + E\left[H(x, X_n)^2 \Big| x \in \bar{X}_n \right] - 2\bar{d}E\left[H(x, X_n) \Big| x \in \bar{X}_n \right] + \bar{d}^2 \right)$$

$$= \frac{1}{2} \left(\bar{d}^2 + E\left[H(x, X_n)^2 \Big| x \in \bar{X}_n \right] - 2\bar{d} \times (2\bar{d}) + \bar{d}^2 \right)$$

$$= \frac{1}{2} \left(E\left[H(x, X_n)^2 \Big| x \in \bar{X}_n \right] - 2\bar{d}^2 \right)$$

Finally, we have

$$C_{FD} = \frac{1}{2^n} \sum_{x \in \{0,1\}^n} (g(x) - \bar{g})(H(x, N(x)) - \bar{d})$$

$$= \frac{1}{2^n} \left(\sum_{x \in X_n} \left(1 - \frac{1}{2} \right)(0 - \bar{d}) + \sum_{x \in \bar{X}_n} \left(0 - \frac{1}{2} \right)(H(x, N(x)) - \bar{d}) \right)$$

$$= \left(-\frac{1}{2} \right) \left(\frac{1}{2}\bar{d} + \frac{1}{2^n} \sum_{x \in \bar{X}_n} H(x, N(x)) - \frac{1}{2^n} \sum_{x \in \bar{X}_n} \bar{d} \right)$$

$$= \left(-\frac{1}{2} \right) \left(\frac{1}{2^n} \sum_{x \in \bar{X}_n} H(x, N(x)) \right)$$

$$= \left(-\frac{1}{2} \right) \left(\frac{1}{2}E[H(x, N(x))|x \in \bar{X}_n] \right)$$

$$= -\frac{\bar{d}}{2}$$

In the following subsections we apply these generic results to our three encoding functions: Parity, Truth Table and Majority.

5.2 *fdc* Under Parity

Let us start with the Parity encoding. The bit strings in $\bar{X}_n$ have all odd parity. Therefore, they can be turned into even-parity global optima by a single bit flip. That is, their Hamming distance from a global optimum is always 1, whereby $E[H(x, X_n)|x \in \bar{X}_n] = 1$. So, from Equation (2) we obtain

$$\bar{d} = \frac{1}{2}.$$

From this, it follows that

$$C_{FD} = -\frac{1}{4}.$$

We also have that $E\left[H(x, X_n)^2 \big| x \in \bar{X}_n\right] = 1$. So,

$$\sigma_D^2 = \frac{1}{2}\left(1 - 2 \times \frac{1}{4}\right) = \frac{1}{4}.$$

Therefore, the fitness distance correlation for OneMax under the Parity encoding is

$$r = \frac{-\frac{1}{4}}{\sqrt{\frac{1}{4}}\sqrt{\frac{1}{4}}} = -1$$

That is, the *fdc* of OneMax is unaffected by the presence of bitwise neutrality under Parity encoding, irrespective of the number of bits (n) one uses.

5.3   *fdc* Under Truth Table

Let us now consider the Truth Table encoding. In order to apply Equation (2) we need to compute $E[H(x, X_n)|x \in \bar{X}_n]$. To do this we will treat $H(x, X_n)$ as a stochastic variable, compute its probability distribution and then make use of the definition of expected value. Let us call $p(d)$ the probability that $H(x, X_n) = d$ for a randomly chosen $x \in \bar{X}_n$.

Let us choose uniformly at random an $x \in \bar{X}_n$ and then choose randomly one of the Hamming-1 neighbours, x', of x. Because the entries of the truth table are randomly assigned, the probability that $x' \in X_n$ is $\frac{1}{2}$. Note that $p(1)$ is the probability that at least one neighbour of x is a member of X_n. Since x has n neighbours and each neighbour's membership of X_n is a Bernoulli trial, we have that

$$p(1) = 1 - \left(\frac{1}{2}\right)^n.$$

So, as n grows, $p(1)$ rapidly approaches 1.

Let us now focus on $p(2)$. This can be seen as the probability of a joint event, i.e., none of the Hamming-1 neighbours of a randomly chosen $x \in \bar{X}_n$ is a member of X_n, but at least one of its Hamming-2 neighbours is. We treat these two events as independent.[1] We know that the probability of the first event is just $1 - p(1) = \left(\frac{1}{2}\right)^n$ and we compute the probability of the second as one minus the probability that none of the Hamming-2 neighbours of x is in X_n. Since there are $\binom{n}{2}$ such neighbours and the probability of each being in X_n is $\frac{1}{2}$, the probability that none of the Hamming-2 neighbours of x is in X_n is $1 - \left(\frac{1}{2}\right)^{\binom{n}{2}}$. Putting everything together we then get

$$p(2) = \left(\frac{1}{2}\right)^n \left(1 - \left(\frac{1}{2}\right)^{\binom{n}{2}}\right)$$

---

[1] This is an approximation, but its accuracy rapidly improves with n. So, our calculations are already very accurate for $n \geq 3$.

Similarly, we get

$$p(3) = \left(\frac{1}{2}\right)^{n+\binom{n}{2}} \left(1 - \left(\frac{1}{2}\right)^{\binom{n}{3}}\right)$$

and, more generally,

$$p(d) = \left(\frac{1}{2}\right)^{\sum_{k=1}^{d-1}\binom{n}{k}} \left(1 - \left(\frac{1}{2}\right)^{\binom{n}{d}}\right).$$

We are now in a position to compute

$$E[H(x, X_n)|x \in \bar{X}_n] = \sum_{d=1}^{n} d \cdot p(d). \tag{3}$$

Note that $p(d)$ is a very rapidly decreasing function. For example, for $n = 4$ we have $p(1) = 0.93750$, $p(2) = 0.061523$, $p(3) = 0.00091553$ and $p(4) = 0.000030518$. Furthermore, as n increases more and more of the probability mass accumulates onto $p(1)$, effectively leading to only $p(1)$ and $p(2)$ having any relevance in the calculation in Equation (3). So, we can write

$$E[H(x, X_n)|x \in \bar{X}_n] \approx p(1) + 2p(2) = 1 + \left(\frac{1}{2}\right)^n - 2\left(\frac{1}{2}\right)^{n+\binom{n}{2}},$$

which makes it clear that for the Truth Table encoding $E[H(x, X_n)|x \in \bar{X}_n] \approx 1 + 2^{-n}$. For example, for $n = 3, 4, 5, 6$, and computing $E[H(x, X_n)|x \in \bar{X}_n]$ using Equation (3) we obtain the values 1.11719, 1.06342, 1.03128, 1.01563, respectively. So, under the Truth Table encoding

$$\bar{d} \approx \frac{1}{2} + 2^{-n-1}$$

From this, it follows that

$$C_{FD} = -\frac{1}{4} - 2^{-n-2}.$$

Using a similar approach we compute

$$E\left[H(x, X_n)^2 \middle| x \in \bar{X}_n\right] = \sum_{d=1}^{n} d^2 \cdot p(d)$$

$$\approx p(1) + 4p(2)$$

$$= 1 + 3\left(\frac{1}{2}\right)^n - 4\left(\frac{1}{2}\right)^{n+\binom{n}{2}}$$

$$\approx 1 + 3 \times 2^{-n}$$

So,

$$\sigma_D^2 \approx \frac{1}{2}\left(1 + 3 \times 2^{-n} - 2 \times \left(\frac{1}{2} + 2^{-n-1}\right)^2\right)$$

$$\approx \frac{1}{2}\left(1 + 3 \times 2^{-n} - \frac{1}{2} - 2^{-n}\right)$$

$$\approx \frac{1}{2}\left(\frac{1}{2} + 2 \times 2^{-n}\right)$$

$$= \frac{1}{4} + 2^{-n}$$

Therefore, the fitness distance correlation for OneMax under the Truth Table encoding is

$$r = -\frac{\frac{1}{4} + 2^{-n-2}}{\sqrt{\frac{1}{4} + 2^{-n}}\sqrt{\frac{1}{4}}} = -\frac{2^{(-n-1)} + \frac{1}{2}}{\sqrt{2^{-n} + \frac{1}{4}}} \approx -1 + 2^{-n}.$$

That is, for $n \geq 5$ or so, also Truth Table induces a form of neutrality which effectively leaves the fdc/problem difficulty unchanged. For relatively small values of n, however, this encoding makes the OneMax problem harder, albeit to a small degree. In Section 6 and 7 we will use $n \geq 5$, for which Truth Table effectively behaves like Parity.

5.4 fdc Under Majority

Let us now consider the Majority encoding. Again, we start by computing $E[H(x, X_n)|x \in \bar{X}_n]$.

With a Majority encoding with $T = n/2$ and n odd, $\bar{X}_n$ is the class of all strings of length n which have $0, 1, \ldots \lfloor T \rfloor$ bits set to 1. That is, we can naturally describe $\bar{X}_n$ by saying that it is contains all strings with unitation value $u < T$. Given a string in $\bar{X}_n$ having unitation u, we can compute how close this is to X_n just by looking at how many additional 1's would be needed to transform the string into a member of X_n. This number is simply $\lceil T - u \rceil$. Since for each unitation class u we have $\binom{n}{u}$ strings, we can then write

$$E[H(x, X_n)|x \in \bar{X}_n] = \frac{1}{2^{n-1}} \sum_{x \in \bar{X}_n} H(x, X_n) = \frac{1}{2^{n-1}} \sum_{u < T} \binom{n}{u} \times \lceil T - u \rceil.$$

This can be computed numerically. For $T = n/2$, n odd, and small values of n, $E[H(x, X_n)|x \in \bar{X}_n]$ grows approximately as $0.63 + 0.37\sqrt{n}$. So, we have

$$\bar{d} \approx 0.315 + 0.185\sqrt{n}$$

and

$$C_{FD} \approx -0.1575 - 0.0925\sqrt{n}.$$

Using a similar approach we compute

$$E\left[H(x, X_n)^2 \big| x \in \bar{X}_n\right] = \frac{1}{2^{n-1}} \sum_{x \in \bar{X}_n} H(x, X_n)^2$$

$$= \frac{1}{2^{n-1}} \sum_{u < T} \binom{n}{u} \times \lceil T - u \rceil^2$$

$$\approx 0.725 + 0.334 \times n$$

for small values of n. So,

$$\sigma_D^2 \approx \frac{1}{2}\left(0.725 + 0.334 \times n - 2\left(0.315 + 0.185\sqrt{n}\right)^2\right)$$

$$\approx 0.133\,n - 0.117\,\sqrt{n} + 0.263$$

Therefore, the fitness distance correlation for OneMax under the Majority encoding is

$$r \approx -\frac{0.315 + 0.185\sqrt{n}}{\sqrt{0.133\,n - 0.117\,\sqrt{n} + 0.263}}.$$

So, in this case there is a much more marked effect of the encoding on the difficulty of a problem with the *fdc* progressively increasing (from the original value of -1) when n increases. For example, for $n = 3, 5, 7, 9, 11$ we obtain *fdc* values of approximately -0.9376, -0.8926, -0.8554, -0.8261 and -0.8028, respectively.

Naturally, theoretical *fdc* calculations could be performed also for the Multimodal problem generator and the Trap function in the presence of bitwise neutrality, although for these functions one could not use Jones' result [19, Appendix D]. We do not report these calculations. However, based on our results with OneMax and the results in [32], it is easy to understand that the Parity and Truth Table encodings have no or limited influence on the difficulty of the Trap and Multimodal functions. However, we should expect Majority to make these problems easier.

6   Phenotypic Mutation Rates

The analysis based on *fdc* indicates that the choice of encoding function used to introduce neutrality may be critical in determining whether a problem is made easier or harder by the introduction of neutrality in evolutionary search. However, fitness landscapes and *fdc* effectively neglect to model the fact that the precise distribution of mutants may have an important effect of search behaviour and performance. For example, *fdc* remains the same irrespective of the mutation probability p_{mut}.

So, to better evaluate benefits and drawbacks of neutrality we want to understand what effects different types of neutral encodings have on the way the search proceeds under mutation. In particular we are interested in understanding how genotypic mutations are related to phenotypic mutations, since only phenotypic changes can lead to fitness changes. To do so, we use the notion of *phenotypic mutation rate.*

When the parity encoding is used, the phenotypic mutation rate corresponding to a genotypic mutation rate p_{mut} is given by:

$$pmut_{phenotypic} = \sum_{i=1,3,5,\dots} \binom{n}{i} p_{mut}^i (1 - p_{mut})^{n-i}$$

This is because only an odd number of genotypic bit-flips can produce a phenotypic change.

When the Truth Table encoding is used, the mutation rate at phenotype level is given by:

$$pmut_{phenotypic} = \frac{1 - (1 - p_{mut})^n}{2}$$

This is because there is the potential for a change in phenotypic value whenever we change the row from which we read out the output in the truth table. This happens if at least one genotypic mutation takes place (hence the factor $1 - (1 - p_{mut})^n$). However, not all row changes lead to a flipped phenotypic bit. Because the table is random, this happens only in 50% of the cases (hence the denominator, 2).

The calculation of the phenotypic mutation rates for Majority are more difficult. We can, however, obtain numerical estimates for these very easily. We do this by generating genotypic mutants of individuals using a particular genotypic mutation rate and recording how frequently the mutants are in a different majority class than the original parents.

In Table 1, we show the phenotypic mutation rates when mutation rates at genotype level are 0.01, 0.06 and 0.1 for Parity, Truth Table and Majority. In the case of Majority figures are estimates obtained by generating 10,000 mutants starting from a uniform random population. As we can see, there are conditions in which different encodings produce similar phenotypic mutation rates. This is the case, for instance, for the pairs of numbers in **boldface**, underlined, in *italics* and in sans serif. Note that the Parity and Truth Table (for the values of n used in the table) leave the fitness distance correlation of a problem unchanged, as discussed in the previous section. So, whenever also the phenotypic mutation rates match, we should expect to see similar performance under these two encodings. We will verify this in the next section.

7   Results and Analysis

For OneMax and the other two problems we have used chromosomes of length $\ell = 14$. For the multimodal landscape we have used $P = 400$ peaks. These were distributed in such a way to give the problem deceptive features. Specifically, the highest peak is at position $111\cdots111$ and the second highest peak is at position $000\cdots000$, the remaining peaks are randomly distributed. This last

Table 1. Phenotypic mutation rates when mutation rates at genotype level are 0.01, 0.06 and 0.1

Type of redundancy	$P_{mut} = 0.01$	$P_{mut} = 0.06$	$P_{mut} = 0.1$
Parity (n bits = 5)	0.0480	*0.2361*	0.3362
Parity (n bits = 6)	0.0571	0.2678	0.3689
Parity (n bits = 7)	0.0659	**0.2957**	0.3951
Parity (n bits = 8)	0.0746	0.3202	0.4161
Truth Table (n bits = 5)	0.0245	0.1331	<u>0.2048</u>
Truth Table (n bits = 6)	0.0293	0.1551	*0.2343*
Truth Table (n bits = 7)	0.0340	0.1758	0.2609
Truth Table (n bits = 8)	0.0386	<u>0.1952</u>	**0.2848**
Majority ($n = 5, T = 2.5$)	0.0168	0.0916	0.1530
Majority ($n = 7, T = 3.5$)	0.0204	0.1072	0.1725

Table 2. Parameters

Parameter	Value
Length of the genome	14
Population Size	80
Generations	100
Mutation Rate (per bit)	0.01, 0.06, 0.1
Number of n bits encoded	5, 6, 7, 8
Independent Runs	1,000

feature makes the problem easier than the trap function. For the trap function we used the following parameters: $u_{min} = 13$, $a = 39$, $b = 40$. Figure 2 depicts this trap function. For the three problems we have used a sample size 4,000 to calculate *fdc*.

The experiments were conducted using a GA with fitness proportionate selection and bit-flip mutation. Runs were stopped when the maximum number of generations was reached. The parameters used are given in Table 2.

Let's start by analysing *fdc* for the problems used in our experiments. Table 3 reports *fdc* for a representation without neutrality and for various forms of neutral encoding. As predicted in Section 4, for the three problems, the Parity and Truth Table encodings leave the *fdc* unchanged w.r.t. whatever value it had in the absence of neutrality.[2] On the contrary, as predicted, Majority moves slightly the *fdc* of a problem towards zero, thereby making easy problems harder and hard problems easier. The question now is: will actual search performance be similarly affected?

---

[2] This is not unexpected, since, as discussed in Section 2, the Parity encoding is a case of trivial neutrality (where the evolution of phenotypic bit strings can be modelled without referring to the corresponding genotypes). Also, the Truth Table encoding effectively becomes a case of trivial neutrality for sufficiently large n.

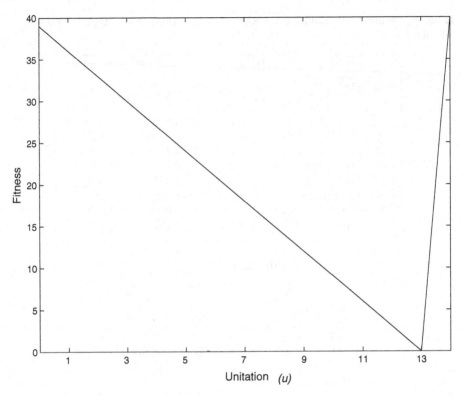

Fig. 2. The trap function used in our experiments ($u_{min} = 13$, $a = 39$, $b = 40$)

In Table 4, we show the average number of generations required to reach the optimum of OneMax and the percentage of successes in finding the optimum measured in 1,000 independent runs of a GA. Let us analyse these results.

When $p_{mut} = 0.01$ we can see a good match between the predictions of *fdc* and problem difficulty. In particular, the Parity and Truth Table encodings show almost exactly the same performance both in terms of percentage of runs where the problem was solved and average number of generations required to solve it. Also, we can see that, as predicted by our *fdc* analysis, the problem is easy and remains easy under all encodings, being solved in almost 100% of cases in all configurations. In addition, we can see that under Majority more generations are required to solve the problem than under Parity and Truth Table, which again confirms the predictions of the *fdc* analysis. There is, however, one element that is unexpected. In the absence of neutrality, runs take longer to find the optimum than with Parity and Truth Table. In fact, they take approximately as long as for Majority.

When $p_{mut} = 0.06$ the situation becomes less clear. Here Parity and Truth Table do not perform identically any more, with Truth Table still being able to solve the problem in almost all runs, while Parity does so only in between 70 and 90% of the cases. This was not predicted by the *fdc* analysis. What is

Table 3. Fitness Distance Correlation estimated for the OneMax problem, the Multimodal Problem generator and the Trap function

Type of redundancy	OneMax Problem	Multimodal Problem	Trap Function
No neutrality	-1	0.5114	0.9979
Parity ($n = 5$)	-1	0.5190	0.9925
Parity ($n = 6$)	-1	0.5190	0.9999
Parity ($n = 7$)	-1	0.5144	0.9999
Parity ($n = 8$)	-1	0.5086	0.9999
Truth Table ($n = 5$)	-0.9999	0.5102	0.9999
Truth Table ($n = 6$)	-1	0.5374	0.9925
Truth Table ($n = 7$)	-1	0.5264	0.9999
Truth Table ($n = 8$)	-0.9999	0.5233	0.9925
Majority ($n = 5, T = 2.5$)	-0.8488	0.4444	0.8434
Majority ($n = 7, T = 3.5$)	-0.8308	0.4471	0.8308

particularly surprising here is that in all cases Parity and Truth Table take longer to solve the problem than Majority and the no-neutrality case. So, Parity and Truth Table effectively make the problem harder, while the other two encodings are still performing approximately the same and their performance seems to be unaffected by the increase in mutation rate. *fdc* analysis also did not predict that performance would vary with n when using the Parity encoding.

These rather confusing trends continue also at the highest genotypic mutation rate, $p_{mut} = 0.1$. Now also the performance with Truth Table varies with n. Furthermore, in the no-neutrality case the problem is now solved in fewer generations than with the Majority encoding.

In summary, it is clear that while *fdc* captures some of the characteristics of a problem in relation to its difficulty for a GA, it does not capture all.

To explain these results one really needs to look at our second descriptor: the phenotypic mutation rates. As one can see in Table 1, when $p_{mut} = 0.01$ the encodings considered induce phenotypic mutation rates in the range 1.5-7.5%. At these mutation rates the GA solves the problem almost equally easily as it does without neutrality.

The more the phenotypic mutation rate is increased, the more the search will be expected to become undirected and random, leading to a worsening of performance. Indeed, when the genotypic mutation rate is increased to 0.06, the Truth Table encoding provides a phenotypic mutation rate which is significantly smaller than for Parity (see Table 1, second column). As expected, in these conditions the performance with Parity is worse than with Truth Table (see Table 4). The phenotypic mutation rates for Majority are even smaller than for Truth Table. So, it is not surprising to see that the GA performs better with Majority than with all other encodings.

When $p_{mut} = 0.1$, the phenotypic mutation rates for all encodings are further increased, leading to an even more undirected search. Note how, in these conditions, the phenotypic mutation rates for Truth Table are similar to those

Table 4. Performance of a mutation-based GA on the OneMax problem. Pairs of numbers in **boldface**, <u>underline</u>, *italics* or sans serif represent situations with almost identical phenotypic mutation rates.

	$p_{mut} = 0.01$		$p_{mut} = 0.06$		$p_{mut} = 0.1$	
	Avr. Gen	% Suc.	Avr. Gen	% Suc.	Avr. Gen	% Suc.
No neutrality	21.35	100%	14.39	100%	16.58	100%
Parity ($n = 5$)	14.55	100%	36.06	*90.1%*	44.02	62.7%
Parity ($n = 6$)	14.46	100%	38.38	82.6%	45.14	54.4%
Parity ($n = 7$)	14.49	100%	40.09	**73.3%**	42.12	49.7%
Parity ($n = 8$)	15.06	100%	43.26	68.2%	44.56	47.6%
Truth Table ($n = 5$)	16.63	99.9%	20.02	99.5%	29.21	<u>95.0%</u>
Truth Table ($n = 6$)	16.89	100%	22.87	99.4%	33.14	*90.5%*
Truth Table ($n = 7$)	15.89	100%	24.41	97.5%	35.49	84.5%
Truth Table ($n = 8$)	15.01	100%	28.16	<u>97.4%</u>	38.89	**78.8%**
Majority ($n = 5, T = 2.5$)	23.39	99.8%	17.26	99.7%	22.08	99.3%
Majority ($n = 7, T = 3.5$)	23.51	99.8%	17.93	100%	22.50	98.6%

observer at $p_{mut} = 0.06$ for Parity, and how performance is similar for these two cases (see Table 4). Similar phenotypic mutation rates and similar performance can also be observed for Majority (at $p_{mut} = 0.1$) and Truth Table (at $p_{mut} = 0.06$). At a mutation rate of 0.1, Parity presents high phenotypic mutation rates, reaching 41.6% in the case $n = 8$. In these conditions the search is almost random and so performance is poor.

Increasing further the genotypic mutation rate will lead the Parity and Truth Table encodings near a phenotypic mutation rate of 50%. There the search is effectively a random search. We do 8,000 trials (80 individuals for 100 generations) in each run, which represent 48.82% of the search space size, 2^ℓ, since $\ell = 14$. However, because of resampling we should only expect to find the optimum with probability 38.3%. (This can be computed using the theory for the coupon collector problem, see [11].) This is the limit performance for high mutation rates.

Let us now consider our second problem: the multimodal problem. For this problem, we tuned the parameters in such a way to make the problem hard, but still easier than the trap problem. Again, at the lowest mutation rate, the predictions of *fdc* are roughly correct: the problem is hard (*fdc* > 0) and remains hard irrespective of the encoding used and Parity and Truth Table lead to the same level of difficulty. Again, however, at the higher mutation rates the situation becomes rather more confusing, with Parity showing improved performance over the other encodings and a dependency of performance on n. Effectively, we can observe the opposite effects as in the OneMax problem. However, the confusion again disappears if we look at the mutation rates corresponding to each encoding.

Finally, let us consider the Trap problem. For this problem, the bigger the value of the slope-change location u_{min}, the harder the problem. In our experiments we chose $\ell = 14$ and $u_{min} = 13$ and, so, the problem is very hard. The behaviour of the evolutionary search in this problem is a mirror image of

Table 5. Performance of a GA on the Multimodal function. Pairs of numbers in **bold-face**, <u>underline</u>, *italics* or sans serif represent situations with almost identical phenotypic mutation rates.

	$p_{mut} = 0.01$		$p_{mut} = 0.06$		$p_{mut} = 0.1$	
	Avr. Gen	% Suc.	Avr. Gen	% Suc.	Avr. Gen	% Suc.
No neutrality	8.56	3.2%	5.22	2.7%	11.54	1.9%
Parity ($n = 5$)	5.61	3.4%	41.2	*5.8%*	44.07	14.2%
Parity ($n = 6$)	4.76	3.4%	45.27	7.2%	50.41	19.4%
Parity ($n = 7$)	2.80	2.1%	44.41	**9.9%**	46.31	24.6%
Parity ($n = 8$)	4.85	2.1%	42.14	12.7%	46.94	23.2%
Truth Table ($n = 5$)	6.41	3.6%	15.86	2.5%	34.11	<u>3.5%</u>
Truth Table ($n = 6$)	8.18	2.5%	20.27	2.2%	34.32	*4.8%*
Truth Table ($n = 7$)	6.59	2.6%	24.07	3.1%	44.44	5.6%
Truth Table ($n = 8$)	4.95	3.6%	19.10	<u>3.2%</u>	33.03	**7.9%**
Majority ($n = 5, T = 2.5$)	11.41	2.0%	23.6	1.4%	15.62	1.9%
Majority ($n = 7, T = 3.5$)	9.76	2.3%	9.44	2.2%	25.42	2.4%

that observed on the OneMax problem (see Table 6). Again, we can see how *fdc* makes reasonably good predictions of relative difficulty under different encodings when $p_{mut} = 0.01$, but that the picture becomes less and less clear as p_{mut} increases. However, again, we can explain performance differences easily by looking at phenotypic mutation rates. In this case, because the problem is deceptive, the more random the search is, the more likely the global optimum is found. So, performance improves as the phenotypic mutation rate increases.

Table 6. Performance of a GA on the Trap function. Pairs of numbers in **boldface**, <u>underline</u>, *italics* or sans serif represent situations with almost identical phenotypic mutation rates.

	$p_{mut} = 0.01$		$p_{mut} = 0.06$		$p_{mut} = 0.1$	
	Avr. Gen	% Suc.	Avr. Gen	% Suc.	Avr. Gen	% Suc.
No neutrality	0.6	0.3%	7.2	0.7%	4.55	0.7%
Parity ($n = 5$)	1	0.5%	47.77	*10.4%*	44.85	22.0%
Parity ($n = 6$)	1	0.8%	45.96	15.6%	44.73	23.8%
Parity ($n = 7$)	1	0.6%	48.62	**15.4%**	46.82	32.0%
Parity ($n = 8$)	13.57	0.7%	46.27	20.2%	46.69	31.5%
Truth Table ($n = 5$)	1	0.7%	13.05	1.4%	41.49	<u>6.3%</u>
Truth Table ($n = 6$)	1.25	0.6%	35.16	2.1%	47.19	*7.8%*
Truth Table ($n = 7$)	1	0.1%	32.36	3.5%	47.32	10.9%
Truth Table ($n = 8$)	1	0.9%	34.44	<u>4.8%</u>	58.54	**13.0%**
Majority ($n = 5, T = 2.5$)	1	1.1%	4.4	1.2%	19.91	2.3%
Majority ($n = 7, T = 3.5$)	1	0.5%	1.16	0.6%	28.15	1.9%

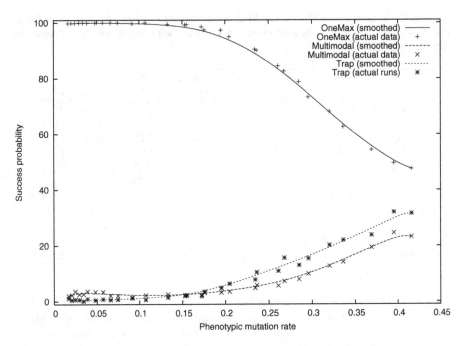

Fig. 3. Success probability vs. phenotypic mutation rates for the OneMax, Multimodal and Trap fitness functions. The points are obtained by combining the entries in Table 1 with those in Tables 4, 5 and 6.

The importance of considering the phenotypic mutation rates instead of the classical genotypic mutation rates is shown in Figures 3 and 4. These figures simply plot the success probabilities reported in Tables 4, 5 and 6 against the corresponding genotypic and phenotypic mutation rates, totally ignoring distinctions between encodings. It is clear how the data strongly correlate with the phenotypic mutation rates, while they correlate much more weakly with the genotypic mutation rates. Note, for example, how the fairly ordinary genotypic mutation rate of 0.1 leads the GA to perform very close to the random search limit (38.3%). We believe this is one of the reasons why so much confusion is present in the EC literature on neutrality.

From these results, it is apparent that *fdc* roughly provides an indication of difficulty, but also that in order to obtain more accurate information one needs to consider how the chosen representation translates genotypic mutation rates into phenotypic mutation rates. With this information in hand, one should then expect to see that for problems with negative *fdc*, performance degrades as the phenotypic mutation rate increases, while the opposite happens for problems with positive *fdc*.

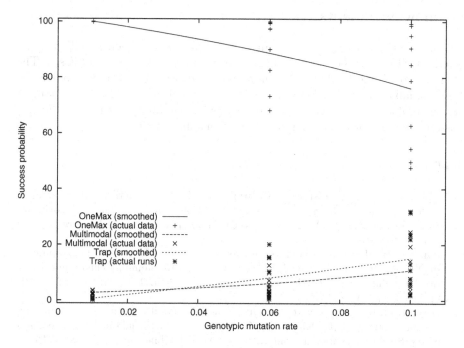

Fig. 4. Success probability vs. genotypic mutation rates for the OneMax, Multimodal and Trap fitness functions. The points are obtained from Tables 4, 5 and 6.

8 Conclusions

In the literature there is contradicting evidence as to whether or not neutrality aids evolutionary search. We believe that, with notable exceptions (e.g., [32]), the confusion often derives from the fact that different researchers use radically different types of neutral encodings and neglect to consider the effect of important parameters such as the rate of application of genetic operators. As we have shown in this paper, small changes in the representation used and search parameters can turn a neutral encoding from being beneficial to being strongly disadvantageous and *vice versa*.

In this paper we considered a form of neutrality induced by a genotype-phenotype map where each phenotypic bit is obtained by transforming a group of genotypic bits via an encoding function. By using explicit calculations for fitness distance correlation, we showed under what conditions a neutral encoding has the potential to induce big changes in problem hardness. We also studied how phenotypic mutation rates change as a function of the genotypic mutation rate for different encodings. We then performed extensive empirical experimentation. We showed that the performance of a GA can change radically with different types of neutrality and mutation rates. However, phenotypic mutation rates and *fdc* allowed us to formulate simple explanations for why this happens.

Acknowledgements

The first author thanks the Mexican Consejo Nacional de Ciencia y Tecnología (CONACyT) for support to pursue graduate studies at University of Essex. The authors would like to thank the anonymous reviewers and Chris Stephens for their valuable comments. Marc Toussaint is particularly thanked for spotting a serious problem in our origianal analysis, for his helpful suggestions and for his costant support throughout the revision of this manuscript.

References

1. Altenberg, L.: Fitness distance correlation analysis: An instructive counterexample. In: Proceedings of the Seventh International Conference on Genetic Algorithms, San Francisco, CA, USA, 1997, pp. 57–64. Morgan Kaufmann Publishers Inc. San Francisco (1997)
2. Banzhaf, W., Leier, A.: Evolution on Neutral Networks in Genetic Programming. In: Yu, T., Riolo, R.L., Worzel, B. (eds.) Genetic Programming Theory and Practice III, volume 9 of Genetic Programming, May 12-14, 2005, ch. 9, pp. 207–221. Springer, Ann Arbor (2005)
3. Barnett, L.: Ruggedness and neutrality – the NKp family of fitness landscapes. In: Adami, C., Belew, R.K., Kitano, H., Taylor, C. (eds.) Artificial Life VI: Proceedings of the Sixth International Conference on Artificial Life, pp. 18–27. MIT Press, Cambridge, MA (1998)
4. Beyer, H.: The Theory of Evolution Strategies. Springer, Heidelberg (2001)
5. Chow, R.: Effects of Phenotypic Feedback and the Coupling of Genotypic and Phenotypic Spaces in Genetic Searchers. In: Proceedings of the 2004 IEEE Congress on Evolutionary Computation, Portland, Oregon, June 20-23, 2004, pp. 242–249. IEEE Press, New York (2004)
6. Clergue, M., Collard, P., Tomassini, M., Vanneschi, L.: Fitness distance correlation and problem difficulty for genetic programming. In: Langdon, W.B., Cantú-Paz, E., Mathias, K., Roy, R., Davis, D., Poli, R., Balakrishnan, K., Honavar, V., Rudolph, G., Wegener, J., Bull, L., Potter, M.A., Schultz, A.C., Miller, J.F., Burke, E., Jonoska, N. (eds.) GECCO 2002. Proceedings of the Genetic and Evolutionary Computation Conference, New York, July 9-13, 2002, pp. 724–732. Morgan Kaufmann, San Francisco (2002)
7. Collins, M.: Finding needles in haystacks is harder with neutrality. In: B, H.-G., et al. (ed.) GECCO 2005. Proceedings of the 2005 conference on Genetic and evolutionary computation, Washington DC, USA, June 25-29, 2005, vol. 2, pp. 1613–1618. ACM Press, New York (2005)
8. De Jong, K., Potter, M.A., Spears, W.M.: Using Problem Generators to Explore the Effects of Epistasis. In: Back, T. (ed.) Proceedings ICGA 1997: International Conference on Genetic Algorithms, San Francisco, USA, 1997, pp. 338–345. Morgan Kaufmann, San Francisco (1997)
9. Ebner, M., Langguth, P., Albert, J., Schakleton, M., Shipman, R.: On Neutral Networks and Evolvability. In: Proceedings of the 2001 Congress on Evolutionary Computation CEC2001, Seoul, Korea, May 27-30, 2001, pp. 1–8. IEEE Press, New York (2001)
10. Eiben, A.E., Hinterding, R., Michalewicz, Z.: Parameter control in evolutionary algorithms. IEEE Trans. Evolutionary Computation 3(2), 124–141 (1999)

11. Feller, W.: An Introduction to Probability Theory and Its Applications. Wiley, Chichester (1968)
12. Fonseca, C., Correia, M.: Developing Redudant Binary Representations for Genetic Search. In: Proceedings of the 2005 IEEE Congress on Evolutionary Computation (CEC-2005), Edinburgh, September 2-4, 2005, pp. 372–379. IEEE, New York (2005)
13. Fontana, W., Schuster, P.: Continuity in evolution: On the nature of transitions. Science 280, 1431–1433 (1998)
14. Galván-López, E., Poli, R.: An Empirical Investigation of How and Why Neutrality Affects Evolutionary Search. In: GECCO 2006. Proceedings of the 2006 conference on Genetic and evolutionary computation, Seattle, WA, USA, July 8-12, 2006, pp. 1149–1156. ACM Press, New York (2006)
15. Goldberg, D.E.: Construction of high-order deceptive functions using low-order walsh coefficients. Ann. Math. Artif. Intell. 5(1), 35–47 (1992)
16. Goldberg, D.E., Deb, K., Horn, J.: Massive multimodality, deception, and genetic algorithms. In: Männer, R., Manderick, B. (eds.) Parallel Problem Solving from Nature, vol. 2, Elsevier Science Publishers, B. V, Amsterdam (1992)
17. Kargupta, K.D.H., Goldberg, D.: Ordering genetic algorithms and deception. In: Männer, R., Manderick, B. (eds.) Parallel Problem Solving from Nature, vol. 2, Elsevier Science Publishers, B. V, Amsterdam (1992)
18. Huynen, M.A.: Exploring phenotype space through neutral evolution. Molecular Evolution 43, 165–169 (1996)
19. Jones, T.: Evolutionary Algorithms, Fitness Landscapes and Search. PhD thesis, University of New Mexico, Albuquerque (1995)
20. Jones, T., Forrest, S.: Fitness distance correlation as a measure of problem difficulty for genetic algorithms. In: Proceedings of the 6th International Conference on Genetic Algorithms, San Francisco, CA, USA, 1995, pp. 184–192. Morgan Kaufmann Publishers Inc. San Francisco (1995)
21. Kennedy, J., Spears, W.M.: Matching Algorithms to Problems: An Experimental Test of the Particle Swarm and Some Genetic Algorithms to the Multimodal Problem Generator. In: Proceedings IEEE International Conference on Evolutionary Computation, Piscataway, NJ, 1998, pp. 78–83. IEEE Press, New York (1998)
22. Kimura, M.: Evolutionary rate at the molecular level. Nature 217, 624–626 (1968)
23. Lobo, G., Lima, C.F.: On the Utility of the Multimodal Problem Generator for Assessing the Performance of Evolutionary algorithms. In: GECCO 2006. Proceedings of the 2006 conference on Genetic and evolutionary computation, Seattle, WA, USA, July 8-12, 2006, pp. 1233–1240. ACM Press, New York (2006)
24. López, E.G., Poli, R.: Some steps towards understanding how neutrality affects evolutionary search. In: Runarsson, T.P., Beyer, H.-G., Burke, E., Merelo-Guervós, J.J., Whitley, L.D., Yao, X. (eds.) Parallel Problem Solving from Nature - PPSN IX. LNCS, vol. 4193, pp. 778–787. Springer, Heidelberg (2006)
25. Miller, J.F., Thomson, P.: Cartesian genetic programming. In: Poli, R., Banzhaf, W., Langdon, W.B., Miller, J., Nordin, P., Fogarty, T.C. (eds.) EuroGP 2000. LNCS, vol. 1802, pp. 121–132. Springer, Heidelberg (2000)
26. Naudts, B., Kallel, L.: A comparison of predictive measures of problem difficulty in evolutionary algorithms. IEEE Transactions Evolutionary Computation 4(1), 1–15 (2000)
27. Quick, R.J., Rayward-Smith, V.J., Smith, G.D.: Fitness distance correlation and ridge functions. In: Eiben, A.E., Bäck, T., Schoenauer, M., Schwefel, H.-P. (eds.) Parallel Problem Solving from Nature - PPSN V. LNCS, vol. 1498, pp. 77–86. Springer, Heidelberg (1998)

28. Rothlauf, F., Goldberg, D.: Redundant representations in evolutionary algorithms. Evolutionary Computation 11(4), 381–415 (2003)
29. Smith, T., Husbands, P., O'Shea, M.: Neutral networks and evolvability with complex genotype-phenotype mapping. In: Kelemen, J., Sosík, P. (eds.) ECAL 2001. LNCS (LNAI), vol. 2159, pp. 272–282. Springer, Heidelberg (2001)
30. Smith, T., Husbands, P., O'Shea, M.: Neutral networks in an evolutionary robotics search space. In: Congress on Evolutionary Computation: CEC 2001, pp. 136–145. IEEE Press, New York (2001)
31. Tomassini, M., Vanneschi, L., Collard, P., Clergue, M.: A study of fitness distance correlation as a difficulty measure in genetic programming. Evolutionary Computation 13(2), 213–239 (Summer 2005)
32. Toussaint, M.: On the evolution of phenotypic exploration distributions. In: Cotta, C., De Jong, K., Poli, R., Rowe, J. (eds.) Foundations of Genetic Algorithms 7 (FOGA 2003), pp. 169–182. Morgan Kaufmann, San Francisco (2003)
33. Toussaint, M., Igel, C.: Neutrality: A necessity for self-adaptation. In: Proceedings of the IEEE Congress on Evolutionary Computation (CEC 2002), pp. 1354–1359 (2002)
34. Vanneschi, L., Tomassini, M., Collard, P., Clergue, M.: Fitness distance correlation in structural mutation genetic programming. In: Ryan, C., Soule, T., Keijzer, M., Tsang, E.P.K., Poli, R., Costa, E. (eds.) EuroGP 2003. LNCS, vol. 2610, pp. 455–464. Springer, Heidelberg (2003)
35. Vassilev, V.K., Miller, J.F.: The advantages of landscape neutrality in digital circuit evolution. In: Miller, J.F., Thompson, A., Thompson, P., Fogarty, T.C. (eds.) ICES 2000. LNCS, vol. 1801, pp. 252–263. Springer, Heidelberg (2000)
36. Weicker, K., Weicker, N.: Burden and Benefits of Redundancy. In: Martin, W., Spears, W. (eds.) Foundations of Genetic Algorithms 6, pp. 313–333. Morgan Kaufmann, San Francisco (2000)
37. Yu, T., Miller, J.: Neutrality and the evolvability of boolean function landscape. In: Miller, J., Tomassini, M., Lanzi, P.L., Ryan, C., Tetamanzi, A.G.B., Langdon, W.B. (eds.) EuroGP 2001. LNCS, vol. 2038, pp. 204–211. Springer, Heidelberg (2001)
38. Yu, T., Miller, J.F.: Needles in haystacks are not hard to find with neutrality. In: Foster, J.A., Lutton, E., Miller, J., Ryan, C., Tettamanzi, A.G.B. (eds.) EuroGP 2002. LNCS, vol. 2278, pp. 13–25. Springer, Heidelberg (2002)
39. Yu, T., Miller, J.F.: The role of neutral and adaptive mutation in an evolutionary search on the onemax problem. In: Cantú-Paz, E. (ed.) Late Breaking Papers at the Genetic and Evolutionary Computation Conference (GECCO-2002), New York, NY, July 2002, pp. 512–519. AAAI, Stanford, California, USA (2002)

Continuous Optimisation Theory Made Easy? Finite-Element Models of Evolutionary Strategies, Genetic Algorithms and Particle Swarm Optimizers

R. Poli[1], W.B. Langdon[2], M. Clerc[3], and C.R. Stephens[4]

[1] Department of Computer Science, University of Essex, UK
[2] Department of Mathematical Sciences, University of Essex, UK
[3] Independent Consultant, Groisy, France
[4] Instituto de Ciencias Nucleares, UNAM, México

Abstract. We propose a method to build discrete Markov chain models of continuous stochastic optimisers that can approximate them on arbitrary continuous problems to any precision. We discretise the objective function using a finite element method grid which produces corresponding distinct states in the search algorithm. Iterating the transition matrix gives precise information about the behaviour of the optimiser at each generation, including the probability of it finding the global optima or being deceived. The approach is tested on a (1+1)-ES, a bare bones PSO and a real-valued GA. The predictions are remarkably accurate.

1 Introduction

Markov chains are important in the theoretical analysis of evolutionary algorithms operating on discrete search spaces. So far they have been of little use for EAs searching on continuous spaces. Naturally, Markov chains with continuous state spaces can be defined and powerful general results have been obtained using them [14]. However, the complexity of the calculations involved makes them less than ideal for the detailed theoretical analysis of continuous optimisers. Instead a variety of different tools have been used. Despite these, generally making theoretical progress in the continuous domain is extremely difficult. As a result, while there are a few continuous domain optimisers, such as evolutionary strategies [2], for which we have a reasonably clear mathematical understanding, in most other cases, the reasons why an algorithm works (or does not) are totally unclear. For example, how differential evolution [16] works is considered by most to be a mystery. In other cases, detailed models of only some components of an algorithm are available, as in the case of genetic algorithms applied to continuous functions [12].

Even where substantial theoretical progress has been made, this has virtually always required either working at a highly abstract level or considering in detail very special cases. For example, in evolutionary strategies, most theory has been restricted to the class of sphere functions. While in the case of particle swarm optimisers, the theory available (see for example [13,5,19,18]) assumes: isolated

C.R. Stephens et al. (Eds.): FOGA 2007, LNCS 4436, pp. 165–193, 2007.
© Springer-Verlag Berlin Heidelberg 2007

single individuals, the search stagnates (i.e., no improved solutions are found) and, until very recently, even that there is no randomness. (In general none of these are true.) So, there is a large gap between theory and practice for continuous optimisers.

We suggest an idea which has the potential to radically improve the situation. It is general and can be applied to most continuous optimisers and arbitrary fitness functions. The inspiration has come from the Finite Element Method. FEM has been very successfully used to model continuous systems in a variety of disciplines [3]. It divides continuous systems into elements. These are sufficiently small that the behaviour of each can safely be modelled by a numerically simple function and so the whole system is accurately modelled by simply combining all its elements. Naturally, the accuracy of the results depends on the resolution of the mesh of elements used (which can be different for different parts of the system). When the mesh is fine enough, the analysis can be extremely accurate.

We discretise the system (in our case, the optimisation algorithm and the fitness function) and then study the dynamics of the discretised system (see Section 2). The new system is in one of a finite number of states. Crucially we will assume the optimisation algorithm's future behaviour can be captured by the current state. This is true of most evolutionary algorithms, which only depend on the current population and not on older populations. Hence we can model the new system as a Markov chain. By studying the chain we can then learn about the behaviour of the original (continuous) system. As we will see there is a notion of resolution. If the discretisation mesh is chosen appropriately, the accuracy with which the chain models the continuous system, over many generations, sometimes even for quite coarse grids can be remarkable.

This gives us an effective general technique to produce discrete Markov chain models of continuous stochastic optimisers that can approximate them on continuous problems to arbitrary precision and for arbitrary fitness functions. The model is complete and includes the ability to estimate arbitrary statistics, such as the evolution of average fitness, best fitness and population diversity. In particular, it is very easy to estimate the probability of an optimiser finding global optima or being deceived.

Our objective is to introduce the idea and to provide a proof of concept for it. So, we will apply the approach to a small, but diverse set of optimisers – (1+1) Evolutionary Strategies (Section 3), Particle Swarm Optimisers (Section 4) and real-valued Genetic Algorithms(Section 5) – and show how easily we can estimate important properties such as the probability of finding the global optimum and the expected runtime of each algorithm (Section 6). We will consider four different problems and will compare the behaviour of the resulting chains with actual runs (Section 7). We postpone the analysis and comparison of the resulting models. We draw some conclusions in Section 8.

2   Discretisation

We can obtain a discrete model of a continuous optimiser in two ways. Firstly, we can perform a formal FEM *approximation of the exact Markov chain representing the optimiser* over continuous search/state-spaces. We illustrate this approach in Section 2.1. Secondly, we can construct a discrete approximation of the optimiser and then obtain an *exact model for an approximation of the optimiser* as shown in Section 2.2. As we will see, when piecewise constant functions are used in FEM, the two approaches lead to the same type of model: a discrete Markov chain.

2.1 FEM Approximation of Exact Markov Chain

Rudolph's EA model. Let us recall the key elements of the generic model of EA presented by Rudolph in [14]. In this model the EA is seen as a homogeneous Markov chain $(X_t : t \geq 0)$ on a probability space $(\Phi, \mathcal{F}, \mathsf{P})$ with image space $(E, \mathcal{A})$, where Φ is the set of outcomes, $\mathcal{F}$ is the set of events (subsets of Φ) and P is a probability measure. Formally $\mathcal{F}$ must be a σ-algebra over Φ, i.e., it must be closed under complementation and countable unions of its members, and $\mathsf{P} : \mathcal{F} \to [0, 1]$ must be a measure and $\mathsf{P}(\Phi) = 1$. The set E is the state space for the system, while $\mathcal{A}$ is a σ-algebra over E. Since an EA consists of a population of N individuals represented by the N-tuple $(x_1, \cdots, x_N)$, where the x_i belong to some domain $\mathcal{M}$ (e.g., $\mathcal{M} = \mathbb{R}$) for $i = 1, \cdots, N$, typically the state space is $E = \mathcal{M}^N$, but there are more complex cases.

In Rudolph's EA model the probabilistic modifications on the population caused by the genetic operators are represented by a stochastic kernel $\mathsf{K}(.,.)$. The map $\mathsf{K} : E \times \mathcal{A} \to [0, 1]$ is termed a Markovian kernel for the chain if $\mathsf{K}(., A)$ is measurable for any fixed set $A \in \mathcal{A}$ and $\mathsf{K}(x, .)$ is a probability measure on $(E, \mathcal{A})$ for any fixed state $x \in E$. In particular, $\mathsf{K}(x_t, A) = \mathsf{P}\{X_{t+1} \in A | X_t = x_t\}$.

The t-th iteration of the Markovian kernel given by

$$\mathsf{K}^{(t)}(x, A) = \begin{cases} \mathsf{K}(x, A) & \text{if } t = 1, \\ \int_E \mathsf{K}^{(t-1)}(y, A)\mathsf{K}(x, dy) & \text{if } t > 1, \end{cases} \tag{1}$$

describes the probability of the EA's state being in some set $A \subseteq E$ within t steps when starting from the state $x \in E$, i.e., $\mathsf{K}^{(t)}(x, A) = \mathsf{P}\{X_t \in A | X_0 = x\}$.

Let $\pi(.)$ denote the initial distribution over subsets A of $\mathcal{A}$, e.g., the probability distribution for the initial population at step $t = 0$. Then

$$\mathsf{P}\{X_t \in A\} = \begin{cases} \pi(A) & \text{if } t = 0, \\ \int_E \mathsf{K}^{(t)}(y, A)\pi(dy) & \text{if } t > 0. \end{cases} \tag{2}$$

FEM applied to Rudolph's model. The starting point of FEM is the definition of a mesh and the assumption that the solution to the problem can be expressed as a piecewise linear, quadratic, or higher order function over the mesh. There is no limitation as to the simplicity (or complexity) of the elements. They can even be constant. This is the type of elements we will use here, although one could extend the results to the case of more sophisticated elements.

In the case of an EA or other stochastic optimiser exploring a continuous search/state space, the function $\mathsf{P}\{X_t \in A\}$ in (2) provides a full probabilistic description of the system. This is, however, clearly a function of the kernel

$K^{(t)}(x, A)$ in (1). We will therefore take the latter as the solution of our problem that we will represent by finite elements.

Let us assume that E is divided up into n disjoint sets E_i such that $E = \bigcup_i E_i$. These represent our mesh. We assume that the family of functions $K^{(t)}(x, A)$ is *piecewise constant* on each element, i.e., $\forall t, \forall i, \forall x', x'' \in E_i, \forall j : K^{(t)}(x', E_j) = K^{(t)}(x'', E_j)$.

Let us now focus on A's which are obtained as the union of *some* E_i, i.e., $A = \bigcup_{i \in I} E_i$, where $I \subseteq \{1, \cdots, n\}$. Then we can write $P\{X_t \in A\} = \sum_{i \in I} P\{X_t \in E_i\}$. We can, therefore, focus our attention on the quantities $P\{X_t \in E_i\}$. For $t > 0$, from (2) we obtain

$$P\{X_t \in E_i\} = \int_E K^{(t)}(y, E_i)\pi(dy)$$

$$= \sum_j \int_{E_j} K^{(t)}(y, E_i)\pi(dy)$$

$$= \sum_j \int_{E_j} K^{(t)}(y_j, E_i)\pi(dy)$$

$$= \sum_j K^{(t)}(y_j, E_i)\pi(E_j)$$

where y_j is any representative element of the set E_j (e.g., its centroid, if the set is compact). From (1) we obtain

$$K^{(t)}(y_j, E_i) = \int_E K^{(t-1)}(y, E_i)K(y_j, dy)$$

$$= \sum_n \int_{E_n} K^{(t-1)}(y, E_i)K(y_j, dy)$$

$$= \sum_n \int_{E_n} K^{(t-1)}(y_n, E_i)K(y_j, dy)$$

$$= \sum_n K^{(t-1)}(y_n, E_i) \int_{E_n} K(y_j, dy)$$

$$= \sum_n K^{(t-1)}(y_n, E_i)K(y_j, E_n)$$

If M is a matrix with elements $m_{ij} = K(y_j, E_i)$ we have that $K^{(t)}(y_j, E_i)$ is the (i, j)-th of M^t and $P\{X_t \in E_i\}$ is the i-th element of $M^t p$, where p is a vector whose elements are $\pi(E_i)$. That is, *the FEM approximation with order-0 elements to Rudolph's exact EA model is an ordinary discrete Markov chain.*

2.2   Exact Markov Model of Approximate Optimiser

An alternative to using FEM is to first obtain a discrete optimiser whose behaviour strongly resembles the behaviour of the original (continuous) optimiser, and then use standard-type Markov chain theory to model such an optimiser.

This approach effectively stands to the previous as the finite difference method (FDM) stands to FEM. FDM is a method for integrating differential equations. The difference between FEM and FDM is that, while in FEM one approximates the solution to a problem, in FDM one discretises the equations of motion of the system. However, it is well known that in certain conditions, e.g., when using piecewise constant functions like we did in Section 2.1, the two methods coincide. Because of its simplicity in the remainder of the paper we will use the second approach.

Fitness Function f Discretisation. We partition a continuous N-dimensional search space Ω into a finite number (n) of compact non-overlapping sub-domains Ω_i. We give each sub-domain Ω_i a fitness value f_i, which can be computed as the mean of f over Ω_i, i.e., $f_i = \int_{\Omega_i} f(x)dx / \int_{\Omega_i} dx$, or simply $f_i = f(x_{c_i})$ where x_{c_i} is the centroid of cell i, i.e., $x_{c_i} = \int_{\Omega_i} xdx / \int_{\Omega_i} dx$. We will call the pair $S_i = (\Omega_i, f_i)$ a *plateau*. So, effectively we turn our continuous fitness function into a piecewise-constant function, which looks like a multidimensional histogram.

If, for example, we consider the case where Ω is a N-dimensional cube which we partition using a regular grid of hypercubic cells, then we can represent each sub-domain as

$$\Omega_i = [x_{c_{i_1}} - r, x_{c_{i_1}} + r] \times [x_{c_{i_2}} - r, x_{c_{i_2}} + r] \times \cdots \times [x_{c_{i_N}} - r, x_{c_{i_N}} + r] \quad (3)$$

where r is the cell "radius" and $x_{c_{i_j}}$ is the j-th component of a lattice point x_{c_i} (the centroid of each sub-domain). So, when r is known and fixed, we can simply represent each plateau using its centroid and fitness value. That is $S_i = (x_{c_i}, f_i)$. Figure 1 shows two one-dimensional fitness functions and two corresponding piecewise-constant functions obtained by discretising them with $r = 0.25$.

Naturally, the choice of the mesh is crucial in determining the accuracy of the resulting model. Clearly, the finer the grid, the more accurate the results. However, also the method with which the fitness of plateaus is computed is important. When these are computed with $f_i = f(x_{c_i})$, as we will do in the rest of the paper, we run the risk of missing important landscape features of sub-element size. This is less likely when f_i is mean of f over Ω_i. The disadvantage of this method is that one needs to compute integrals of the fitness function.

Algorithm Discretisation. Most optimisers store information about one or more points in Ω which are used to determine which areas of the search space to sample next. Let us assume that there are P such points, which we will call a population. The population is the state of the optimiser. To discretise the optimiser we need to discretise its population s so that instead of taking continuous values it can only take a finite number of states. We use the same discretisation mesh $\{\Omega_i\}$ as for the fitness function. So, in the discretised optimiser the population s is in one of the states $\{x_{c_i}\}^P$.

Some optimisers use additional variables and parameters to control the search, and these are often adapted dynamically. E.g., an ES may change the mutation strength, while the velocities of the particles change in a PSO. When these quantities adapt during the search they are part of the state of the algorithm. Hence they must also be discretised but the lattice used will depend upon the algorithm.

We should note that both discretisation methods discussed above assume that the search is contained within a finite domain Ω (typically a multidimensional box). This is what most problems require and what many optimisers do. However, some optimisers have unbounded state spaces. So, one cannot be sure whether a certain state variable will stay permanently in pre-defined bounds. There are many strategies to circumvent this problem. One could, for example, use boundary elements of infinite size (but with an artificial, finite, centroid), or use mapping/squashing functions to map an infinite space into a new, finite one. These and other strategies put forward in the FEM community (e.g., [1,7,6,17]), however, are beyond the scope of this article.

3   Evolutionary Strategy Model

To start with, let us consider the simplest possible evolutionary strategy: a (1+1)-ES with Gaussian mutations but *without* adaptation of the mutation standard deviation σ.

Naturally, at any given time the only member of the population, x_p, will be located in some sub-domain Ω_i. After discretisation, x_p can only take one of a discrete set of values, namely $x_p = x_{c_k}$ for k in $\{1, \cdots, n\}$. So, our (1+1)-ES can only be in one of n states. We will indicate the state of the ES with an integer s.

Our objective is to model this simple ES as a Markov chain with states of this form. What we need to do is to compute the state transition matrix $M = (m_{ij})$, where m_{ij} is the probability of the ES moving from state i to state j at the next iteration. When M is available, we can compute the probability distribution π_t of the discretised ES being in any particular state at generation t, given its state probability distribution at the start, π_0, from $\pi_t = M^t \pi_0$.

Let $p(x|x_p)$ be the sampling probability density function when the parent is x_p. Normally in an (1+1)-ES random numbers are chosen independently for each dimension when computing mutants. In our discretised ES we do the same thing. So, we have separability of p. That is, p is given by a product of independent probability distributions for each separate dimension:

$$p(x|x_p) = \prod_{j=1}^{N} p(x_j|x_{p_j})$$

where x_j and x_{p_j} are the j-th components of the vectors x and x_p, respectively.[1]

The probability of sampling sub-domain Ω_i is given by

$$\Pr(\Omega_i|x_p) = \int_{\Omega_i} p(x|x_p)\ dx$$

---

[1] Note that, while separability of the sampling distribution makes the model's calculations simpler, it is not a requirement.

So, if sub-domains Ω_i have the product structure shown in Equation 3, we have

$$\Pr(\Omega_i|x_p) = \prod_j \Pr([x_{c_{i_j}} - r, x_{c_{i_j}} + r]|x_{p_j}) \tag{4}$$

where

$$\Pr([x_{c_{i_j}} - r, x_{c_{i_j}} + r]|x_{p_j}) = \int_{x_{c_{i_j}} - r}^{x_{c_{i_j}} + r} p(x_j|x_{p_j})\, dx_j. \tag{5}$$

The standard sampling distribution used in the ES is a Gaussian distribution, i.e., $p(x_j|x_{p_j}) = G\left(x_{p_j}, \sigma\right)$ with $G(\mu, \sigma) = \frac{1}{\sqrt{2\pi}\,\sigma}e^{-\frac{(x-\mu)^2}{2\sigma^2}}$. Let erf be the integral of the Gaussian distribution. Therefore,

$$\Pr([x_{c_{i_j}} - r, x_{c_{i_j}} + r]|x_{p_j})$$
$$= \frac{1}{2}\left(\mathrm{erf}\left(\frac{x_{c_{i_j}} + r - x_{p_j}}{\sigma\sqrt{2}}\right) - \mathrm{erf}\left(\frac{x_{c_{i_j}} - r - x_{p_j}}{\sigma\sqrt{2}}\right)\right). \tag{6}$$

Let us now put the sub-domains in order of their fitness so that $f_i \le f_j$ for $i < j$. Since the population can only change if there is a fitness improvement, only certain state transitions can occur. That is, a transition from state s to state s' is possible only if $s \le s'$.

Suppose the parent is in domain k, then the probability of it changing to domain l is given by:

$$\Pr(l|k) = \begin{cases} \Pr(\Omega_l|x_{c_k}) & \text{if } l \text{ and } k \text{ are such that } f_l > f_k \text{ (NB: } f_l > \\ & f_k \implies l > k \text{ but not } vice\ versa), \\ 0 & \text{if } k \ne l \text{ and } f_l \le f_k, \\ 1 - \sum_{l:f_l > f_k} \Pr(l|k) & \text{if } l = k, \text{ to guarantee the conservation of} \\ & \text{probability.} \end{cases}$$

This effectively means that the population remains in domain k if any of the following three conditions is met: (a) the new sample is in Ω_k, (b) the new sample is in an Ω_j (different from Ω_k) with $f_j \le f_k$, or (c) the sample is outside Ω. So, we can then write the state transition probability for the ES as

$$m_{s,s'} = \Pr(s'|s) = \Pr(\Omega_{s'}|x_{c_s})\delta(f_{s'} > f_s) + (1 - \sum_{l:f_l > f_s} \Pr(\Omega_l|x_{c_s}))\delta(s' = s), \tag{7}$$

where $\Pr(\Omega_{s'}|x_{c_s})$ and $\Pr(\Omega_l|x_{c_s})$ can be computed using Equations 4 and 6. The function $\delta(z)$ returns 1 if z is true and 0 otherwise.

As an example, consider a domain $\Omega = [-2, 2] \times [-2, 2] = [-2, 2]^2$, and let us divide it into four squared sub-domains $\Omega_1 = [-2, 0)^2$, $\Omega_2 = [-2, 0) \times [0, 2]$, $\Omega_3 = [0, 2] \times [-2, 0)$, and $\Omega_4 = [0, 2]^2$ of radius $r = 1$. These have centroids $x_{c_1} = (-1, -1)$, $x_{c_2} = (-1, 1)$, $x_{c_3} = (1, -1)$ and $x_{c_4} = (1, 1)$. Let us further assume that the fitness function f takes the following values at the centroids: $f_1 = 1$, $f_2 = 2$, $f_3 = 3$, and $f_4 = 4$. Then by applying the equations above, for $\sigma = 1$, we obtain the transition matrix:

$$M = \begin{pmatrix} 0.9499 & 0.0000 & 0.0000 & 0.0000 \\ 0.0232 & 0.9731 & 0.0000 & 0.0000 \\ 0.0232 & 0.0037 & 0.9768 & 0.0000 \\ 0.0037 & 0.0232 & 0.0232 & 1.0000 \end{pmatrix}.$$

By iterating the corresponding chain one can compute the distribution of states of our discretised fixed-σ ES acting on fitness function f at any generation. However, given the very low resolution chosen, we would not expect the predictions of the chain to exactly reflect the real behaviour of the continuous optimiser. As we will see later, however, with higher resolutions, predictions can be very accurate.

Let us now generalise this model to include a more interesting version of (1+1)-ES: one where σ adapts during evolution. To keep our description simple, we will focus on an adaptive scheme which updates σ at each iteration [2, page 84]. If the offspring produced by mutation is better than its parent (and is in Ω) we increase σ according to the rule $\sigma' = \sigma c$ where c is a suitable constant > 1. If, the offspring is invalid or is not better than the parent, we reduce σ using the rule $\sigma' = \sigma/c$.

Naturally, for this new ES we can still discretise the parent individual using the regular mesh adopted for the fitness function, as we did for the fixed-σ case. However, we will use a non-uniform discretisation for σ. Indeed, it is apparent that σ can only take discrete values already, all σ's being of the form $\sigma = \sigma_0 \cdot c^i$ for some integer i, where σ_0 is the value of σ at generation 0. So, in any finite run of G generations, $\sigma \in \{\sigma_0 \cdot c^{-G}, \sigma_0 \cdot c^{-G+1}, \cdots, \sigma_0 \cdot c^G\}$, that is it can only take $2G+1$ different values. Following standard practice, in our ES we will limit σ so that it never becomes too little or too big. This effectively means that we can use a smaller range $\{\sigma_0 \cdot c^{-Z}, \sigma_0 \cdot c^{-Z+1}, \cdots, \sigma_0 \cdot c^Z\}$, with $Z < G$. So, we can represent the state of the ES with the tuple (s_1, s_2), where $s_1 \in \{1, \cdots, n\}$ represents the position of the parent and $s_2 \in \{-Z, \cdots, Z\}$ gives the mutation σ used to create its child. For the purpose of indexing the elements of the array M, we then convert tuples into natural numbers by using the odometer ordering, whereby $(1, -Z)$ maps to 1, $(1, -Z+1)$ maps to 2, etc. (i.e., $(s_1, s_2) \mapsto (2Z+1)s_1 + s_2 - Z$).

The calculations to compute M for the adaptive ES are based on the application of Equation 7, with minor changes. Firstly, when we compute the probability of a transition from state $s = (s_1, s_2)$ to state $s' = (s_1', s_2')$, we use the σ corresponding to s_2, i.e., $\sigma = \sigma_0 \cdot c^{s_2}$. Secondly, for all state pairs where $s_1 < s_1'$ (there was a fitness improvement) but where $s_2' \neq s_2 + 1$ (σ was *not* increased according to our update rule), we know that $m_{s,s'} = 0$, so we don't apply Equation 7. Likewise, for all state pairs where $s_1 \geq s_1'$ and where $s_2' \neq s_2 - 1$.

4   Particle Swarm Optimisation Model

4.1   Background

Particle Swarm optimisers (PSOs) [8,10] have been with us a few years. However it is fair to say that most work on PSOs has been experimental confirmations of their effectiveness, extensions to new applications or new algorithms. With very few exceptions (e.g., see the dynamical system model in [5,18] or the probabilistic stagnation analysis in[4]), analytical, theoretical and mathematical analysis of them is still relatively unexplored.

In a simple PSO, the swarm consists of a population of identical particles which move across a problem landscape looking for high-fitness regions. The particles have momentum and are accelerated by forces applied to them. The PSO's integration of Newton's laws of motion is discrete and the particles only sample the fitness landscape at discrete time steps. Thus the PSO particles draw samples from the search space only at some points in their trajectories. In the classic PSO, there are two attractive forces. The first pulls the particle towards the best point it personally has ever sampled, whilst the second pulls it towards the best point seen by any particle in its neighbourhood. The strengths of the various forces are randomly controlled. It is the stochastic nature of the PSO which allows it to effectively explore and ensures that the loci of the particles are not closed trajectories. Instead, the particles randomly sample the region nearby and between the particle's own best and the swarm best.

One of the recent advances has been Jim Kennedy's "Bare Bones" PSO (BB-PSO) [9]. This optimiser is inspired by the observation that, at least until a better location in the search space is sampled, the pseudo chaotic particle orbits can be approximated by a fixed probability distribution centred on the point lying halfway between the particle best and the swarm best. Its width is modulated by the distance between them. The exact nature of the distribution is not clear: it is bell shaped like a Gaussian distribution [11] but the tails appear to be heavier, like a Cauchy distribution. The essential "bare bones" PSO, cuts out the integration needed to find each particle's position, and instead draws it from a random distribution. This means we no longer need to track exactly each particle's position and velocity. As with other swarm intelligence techniques, there has been little theoretical work on this essential PSO. The model we are about to present addresses this.

4.2   Model of "bare bones" PSO

Let us consider a fully-connected bare bones PSO to start with. In this PSO the particles have no dynamics, but simply sample the neighbourhood of their personal best and swarm best using a fixed probability density function. This continues until either their personal best or the swarm best is improved. When this happens, the parameters of the sampling distribution are recomputed and the process is restarted.

In the unlikely event that more than one particle's personal-best fitness is the same as the best fitness seen so far by the whole swarm, we assume that swarm leadership is shared. That is, each particle chooses as its swarm best a random individual out of the set of swarm bests.

Naturally, at any given time the personal best for each particle and the swarm best will be located in some sub-domain Ω_i. In a discretised BB-PSO both the particle best x_p and swarm best x_s can only take one of a discrete set of values, namely $x_p = x_{c_k}$ and $x_s = x_{c_j}$ for some j and k in $\{1, \cdots, n\}$. So, the discretised algorithm can only be in a finite set of states. However, we don't need to represent explicitly the swarm best, since the information is implicit in the fitness values f_i associated to each centroid. So, if P is the population size, there are n^P such states – one for each particle's personal best – and we can represent states as P dimensional vectors with integer elements, e.g.

$$s = (s_1, \cdots, s_P).$$

Let us now focus on computing state transition probabilities.

Let $p(x|x_s, x_p)$ be the sampling probability density function when swarm best is x_s and particle best is x_p. The standard sampling distribution used in the BB-PSO is a Gaussian distribution. (Our approach could also be applied to Cauchy or other distributions.) So, we have

$$p(x_j|x_{s_j}, x_{p_j}) = G\left(\frac{x_{s_j} + x_{p_j}}{2}, |x_{s_j} - x_{p_j}|\right).$$

Note that this distribution becomes a Dirac delta function when $x_{s_j} = x_{p_j}$. Normally in PSOs random numbers are chosen independently for each dimension when computing force vectors. In a bare bones PSO we do the same thing. So, again we have separability of p and we can write

$$p(x|x_s, x_p) = \prod_{j=1}^{N} p(x_j|x_{s_j}, x_{p_j})$$

where x_j, x_{s_j}, and x_{p_j} are the j-th components of the vectors x, x_s and x_p, respectively.

Similarly to the ES case, the probability of sampling domain Ω_i is given by the integral of p across Ω_i, and, if sub-domains Ω_i have the product structure shown in Equation 3, we have

$$\Pr(\Omega_i|x_s, x_p) = \prod_j \Pr([x_{c_{i_j}} - r, x_{c_{i_j}} + r]|x_{s_j}, x_{p_j}) = \prod_j \int_{x_{c_{i_j}} - r}^{x_{c_{i_j}} + r} p(x_j|x_{s_j}, x_{p_j}) \, dx_j.$$

$$(8)$$

For a Gaussian sampling distribution we have

$$\Pr([x_{c_{i_j}} - r, x_{c_{i_j}} + r]|x_{s_j}, x_{p_j})$$
$$= \begin{cases} \frac{1}{2}\left(\text{erf}\left(\frac{x_{c_{i_j}} + r - \frac{x_{s_j} + x_{p_j}}{2}}{|x_{s_j} - x_{p_j}|\sqrt{2}}\right) - \text{erf}\left(\frac{x_{c_{i_j}} - r - \frac{x_{s_j} + x_{p_j}}{2}}{|x_{s_j} - x_{p_j}|\sqrt{2}}\right)\right) & \text{if } x_{s_j} \neq x_{p_j}, \\ \delta(x_{p_j} \in [x_{c_{i_j}} - r, x_{c_{i_j}} + r]) & \text{otherwise.} \end{cases}$$

Again, let us order sub-domains so that $f_i \leq f_j$ for $i < j$. Since particle personal bests can only change if there is a fitness improvement, only certain state transitions can occur. That is, a transition from state $s = (s_1, \cdots, s_P)$ to state $s' = (s_1', \cdots, s_P')$ is possible only if $s_1 \leq s_1'$, $s_2 \leq s_2'$, etc. We will denote this by $s \leq s'$.

Let us identify the location of the swarm best for a PSO in a state s. Typically in a fully-connected PSO there is only one particle with the best fitness value, but, within a discretised PSO, it is not uncommon to have more than one. So, in general, we have a set of swarm bests:

$$\mathcal{B}(s) = \bigcup_{i:f(s_i)=f_m(s)} \{s_i\}$$

where $f_m(s) = \max_j f_{s_j}$ and $|\mathcal{B}(s)| \geq 1$. More generally, to allow other communication topologies, we need to talk about sets of neighbourhood bests – one set for each particle. We will denote these sets as $\mathcal{B}(s,i)$, for $i = 1, \cdots, P$.

Let us consider a PSO in state s. In a BB-PSO, at each iteration, the particles sample the search space independently. So, if the i-th particle's best is in plateau k (that is, $s_i = k$), then the probability of it changing to plateau l is given by:

$$\Pr(l|\mathcal{B}(s,i),k) = \begin{cases} \frac{1}{|\mathcal{B}(s,i)|}\sum_{b\in\mathcal{B}(s,i)}\Pr(\Omega_l|x_{c_b},x_{c_k}) & \text{if } l \text{ and } k \text{ are such that } f_l > f_k, \\ 0 & \text{if } k \neq l \text{ and } f_l \leq f_k, \\ 1 - \sum_{l:f_l>f_k}\Pr(l|\mathcal{B}(s,i),k) & \text{if } l = k \text{ (to guarantee the conservation of probability).} \end{cases}$$

Like for the ES case, this effectively means that the particle remains in plateau k if any of the following three conditions is met: (a) the new sample is in Ω_k, (b) the new sample is in an Ω_j (different from Ω_k) with $f_j \leq f_k$, or (c) the sample is outside Ω.

Because of the independence of the particles (over one time step), we can then write the state transition probability for the whole PSO as

$$m_{s,s'} = \prod_i \Pr(s_i'|\mathcal{B}(s,i),s_i)$$

$$= \prod_{i:f_{s_i'}>f_{s_i}} \frac{1}{|\mathcal{B}(s,i)|} \sum_{b\in\mathcal{B}(s,i)} \Pr(\Omega_{s_i'}|x_{c_b},x_{c_{s_i}})$$

$$\times \prod_{i:f_{s_i'}\leq f_{s_i}} \left[\left(1 - \sum_{l:f_l>f_{s_i}} \frac{1}{|\mathcal{B}(s,i)|} \sum_{b\in\mathcal{B}(s,i)} \Pr(\Omega_l|x_{c_b},x_{c_{s_i}})\right)\delta(s_i' = s_i)\right].$$

Naturally, further decompositions can be obtained using Equation 8.

As an example, let us consider again the domain $\Omega = [-2,2]^2$, which we divide into four sub-domains Ω_i of radius $r = 1$, with centroids $x_{c_1} = (-1,-1)$, $x_{c_2} = (-1,1)$, $x_{c_3} = (1,-1)$, $x_{c_4} = (1,1)$ and associated fitness $f_1 = 1$, $f_2 = 2$, $f_3 = 3$, and $f_4 = 4$, respectively.

If the population includes only two particles ($P = 2$), then we have only 16 different states for the PSO: (1,1), (1,2), ..., (4,4). For example, the state $(1,1)$ represents the situation where both particles are in the lowest plateau (so, both are swarm bests); in state $(1, 2)$ one particle is in the lowest plateau, while the other (the swarm best) is in the second lowest plateau; etc.

By applying the previous equations we can then obtain the following transition matrix:

(1,1)	(1,2)	(1,3)	(1,4)	(2,1)	(2,2)	(2,3)	(2,4)	(3,1)	(3,2)	(3,3)	(3,4)	(4,1)	(4,2)	(4,3)	(4,4)	
1	0	0	0	0	0	0	0	0	0	0	0	0	0	0	0	(1,1)
0	0.659	0	0	0	0	0	0	0	0	0	0	0	0	0	0	(1,2)
0	0	0.659	0	0	0	0	0	0	0	0	0	0	0	0	0	(1,3)
0	0	0	0.651	0	0	0	0	0	0	0	0	0	0	0	0	(1,4)
0	0	0	0	0.659	0	0	0	0	0	0	0	0	0	0	0	(2,1)
0	0.341	0	0	0.341	1	0	0	0	0	0	0	0	0	0	0	(2,2)
0	0	0	0	0	0	0.766	0	0	0	0	0	0	0	0	0	(2,3)
0	0	0	0.117	0	0	0	0.659	0	0	0	0	0	0	0	0	(2,4)
0	0	0	0	0	0	0	0	0.659	0	0	0	0	0	0	0	(3,1)
0	0	0	0	0	0	0	0	0	0.766	0	0	0	0	0	0	(3,2)
0	0	0.341	0	0	0	0.117	0	0.341	0.117	1	0	0	0	0	0	(3,3)
0	0	0	0.117	0	0	0	0	0	0.117	0	0.659	0	0	0	0	(3,4)
0	0	0	0	0	0	0	0	0	0	0	0	0.651	0	0	0	(4,1)
0	0	0	0	0	0	0	0	0	0	0	0	0.117	0.659	0	0	(4,2)
0	0	0	0	0	0	0.117	0	0	0	0	0	0.117	0	0.659	0	(4,3)
0	0	0	0.117	0	0	0	0.341	0	0	0	0.341	0.117	0.341	0.341	1	(4,4)

where we have added one extra row and column to more clearly identify states.

5   Real-Valued Genetic Algorithm Model

We consider a simple real-valued GA with finite population, fitness proportionate selection, no mutation, and 100% recombination. Recombination produces the offspring, $o = (o_1, \cdots, o_n)$, by sampling uniformly at random within the hyper-parallelepiped defined by the parents, $p' = (p'_1, \cdots, p'_N)$ and $p'' = (p''_1, \cdots, p''_N)$. That is, $o_i = \rho_i(p''_i - p'_i) + p'_i$, where ρ_i is a uniform random number in $[0, 1]$ for $i = 1, \cdots, N$. We will refer to this type of recombination as blend crossover.

We use the same state representation as for BB-PSO, $s = (s_1, \cdots, s_P)$, except that we interpret each s_i as the position of an individual in the search space, rather than a particle's best.

The (offspring) sampling distribution for parents p' and p'' under blend recombination is

$$p(o|p', p'') = \prod_i p(o_i|p'_i, p''_i)$$

where

$$p(o_i|p'_i, p''_i) = \begin{cases} 1/|p'_i - p''_i| & \text{if } o_i \in [\min(p'_i, p''_i), \max(p'_i, p'_i)], \\ 0 & \text{otherwise.} \end{cases}$$

Note that the sampling distribution becomes a Dirac delta function when $p' = p''$.

As before, the probability of sampling domain Ω_i is given by the integral of p across Ω_i. So, for sub-domains Ω_i as in Equation 3, we have

$$\Pr(\Omega_i|p',p'') = \prod_j \int_{x_{c_{i_j}}-r}^{x_{c_{i_j}}+r} p(o_j|p_j',p_j'')\,do_j \tag{9}$$

where

$$\int_{x_{c_{i_j}}-r}^{x_{c_{i_j}}+r} p(o_j|p_j',p_j'')\,do_j$$

$$= \begin{cases} \max\left(0, \dfrac{\min(x_{c_{i_j}}+r,\max(p_j',p_j''))-\max(x_{c_{i_j}}-r,\min(p_j',p_j''))}{|p_j'-p_j''|}\right) & \text{if } p_j' \neq p_j'', \\ \delta(p_j'' \in [x_{c_{i_j}}-r, x_{c_{i_j}}+r]) & \text{otherwise.} \end{cases}$$

By adding the contributions from all possible pairs of parents (with their selection probabilities) we can now compute the total probability that the offspring will sample domain Ω_i in a particular population $\mathcal{P} = (p_1,\cdots,p_P)$:

$$\Pr(\Omega_i|\mathcal{P}) = \sum_{p'\in\mathcal{P}}\sum_{p''\in\mathcal{P}} \Pr(\Omega_i|p',p'')\phi(p')\phi(p'') \tag{10}$$

where $\phi(x)$ is the selection probability of parent x in population $\mathcal{P}$. For fitness proportionate selection $\phi(x) = f(x)/\sum_{y\in\mathcal{P}} f(y)$.

Naturally, when $\mathcal{P}$ is the population associated to state s, Equation (10) gives us the probability, $\Pr(\Omega_i|s)$, of generating an individual in domain i for a population in state s. Because each individual in a population is generated by an independent Bernoulli trial, we can then trivially compute the Markov chain transition probability from any state s to any state s' as

$$m_{s,s'} = \prod_i \Pr(\Omega_{s_i'}|s).$$

6 Success Probability and Expected Run Time of Continuous Optimisers

As mentioned above, when M is available, we can compute the probability distribution π_t of a discretised continuous optimiser being in any particular state at generation t, given its state probability distribution at the start, π_0, from $\pi_t = M^t\pi_0$. Since for each optimiser we know what π_0 is, to compute the probability with which the element containing the global optimum is visited at a particular generation t, one only needs to add up the appropriate components of the π_t vector. We will informally call this quantity the *success probability*. For example, in an ES with fixed σ we have that the components of π_0 are all $1/n$ and the success probability is simply given by the last component of π_t (assuming domains are ordered by fitness).

We can also estimate the *expected run time* of continuous optimisers by computing the expected waiting time of the corresponding discrete Markov chain to visit a particular target state or set of states J. Following [15, pages 168–170] we have that the *mean passage time* for going from state i to the set of states J, given that it is currently outside the set is given by:

$$\eta_{i,J} = \sum_{j \in J} m_{i,j} + \sum_{k \notin J} m_{i,k}(1 + \eta_{k,J}) \tag{11}$$

where $m_{i,j}$ are the elements of Markov matrix for the system. Simple algebraic manipulations of (11) lead to the following system of simultaneous equations:

$$\eta_{i,J} - \sum_{k \notin J} m_{i,k}\eta_{k,J} = 1 \tag{12}$$

Once solved, we can then compute the *expected waiting time* to reach state J, given a random initial state (described by the distribution $p(.)$) as

$$EWT_J = \sum_{i \notin J} p(i)\eta_{i,J}. \tag{13}$$

If the calculation is applied to an initial distribution where all states are equally likely (the standard initialisation strategy in EAs) and with J being the element containing the global optimum, we have

$$E[\text{runtime}] = \frac{\sum_{i \notin J} \eta_{i,J}}{\text{number of elements} - 1}. \tag{14}$$

Note that this calculation assumes that the algorithm has a way of identifying when the element containing the global optimum is sampled and stops when this happens. Often this is not the case. In this case, (14) should be interpreted as the average first hitting time.

7   Experimental Results

We first apply the Markov chain models described in the previous sections to the two one-dimensional ($N = 1$) fitness functions in Figure 1. These are effectively continuous versions of the onemax problem (Ramp) and the deceptive trap function. The results of these tests are reported in Sections 7.1– 7.3. Then, in Section 7.4, we study two-dimensional problems.

7.1   Evolutionary Strategies

In a series of experiments we applied the model of the $(1+1)$-ES with fixed-σ and compared its behaviour with the behaviour of the real algorithm acting on the Ramp and Deceptive continuous fitness functions. We chose the domain $\Omega = [-0.5, 4.5)$. This was divided into n sub-domains $\Omega_i = [x_{c_i} - r, x_{c_i} + r)$ with

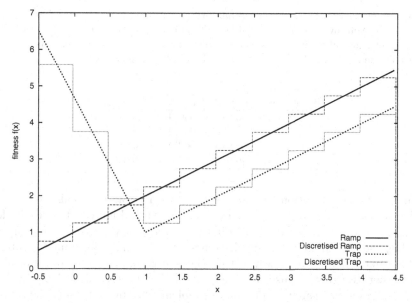

Fig. 1. Ramp and Deceptive test functions and two corresponding piece-wise constant discretisations

equally spaced centroids x_{c_i}. To assess the behaviour of the algorithm in real runs we performed 1,000 independent runs, where, in each run and each generation, we recorded the sub-domain Ω_i occupied by the parent. In particular we were interested in comparing the proportion of runs in which the individual was in the domain containing the global optimum Ω_n vs. the frequency predicted by the Markov chain over a number of generations.

Figure 2 shows the results of the comparison for different values of σ and for the case of $n = 10$ ($r = 0.25$), i.e., where we discretise the evolutionary strategy using only 10 states. As the figure indicates, as long as σ is bigger than the cell width, $2r$, the model predicts the success rate with considerable accuracy throughout our runs (50 generations). When σ is comparable to r, there are errors of up to around 10% in the prediction. Similar accuracies were obtained for the deceptive fitness function (see Figure 3). In all cases, despite its tiny size, the chain was able to predict that Deceptive is harder than Ramp.

To illustrate how one can use our Markov model to study how the computational complexity of an algorithm varies as the parameter σ varies and as a function of the fitness function we also computed (as described in Section 6) the expected first hitting time for the global optimum for the Ramp and Deceptive functions. In this case we used a model with $n = 40$ elements to ensure good accuracy also at small values of σ. Figure 4 shows the results for Ramp. As one can see too small values of σ slow down the march towards the optimum, while too big values make the search excessively random (note, resampling and the rejection of samples outside Ω make the search even slower than pure enumeration which on average would require 20 trials to find the optimum). So, the

Table 1. Comparison between the probability of sampling the optimum domain predicted by the Markov chain for a variable-σ ES and empirical data (averages over 1,000 independent runs) after 50 generations for different discretisation resolutions, n. (NB the size of the optimum domain reduces as the resolution increases.)

Resolution (n)	Ramp		Deceptive	
	Model	Runs	Model	Runs
5	0.973	1.000	0.328	0.378
10	0.998	1.000	0.376	0.388
20	0.998	1.000	0.374	0.381
40	0.988	0.985	0.344	0.349

optimum σ for this function appears to be between 0.5 and 1, as also suggested by the success rates reported in Figure 2. As shown in Figure 5, the results for for Deceptive are radically different. Firstly, for very low values of σ the problem of finding the expected hitting time becomes unstable and so we cannot compute reliable values. It is clear, however, that the search for the global optimum becomes easier as σ grows. This is to be expected. Most evolutionary algorithms do worse than random search on deceptive problems. So, by increasing the search variance, we turn our ES more and more into a random searcher, thereby improving performance (although resampling and the rejection of samples outside Ω prevent performance to ever reaching the pure enumeration limit of 20).

We then considered the (1+1)-ES with variable σ and studied the proportion of runs in which the individual occupied the domain containing the global optimum Ω_n. In our tests we allowed both the discretise algorithm and real one to use a range of 21 different σ's in the range $\{\sigma_0 \cdot c^{-Z}, \sigma_0 \cdot c^{-Z+1}, \cdots, \sigma_0 \cdot c^Z\}$ with $\sigma_0 = 1$, $Z = 10$ and $c = 1.1$. Table 1 shows how the accuracy of the model varies as a function of the number of domains (n). Only the smallest chain, where $n = 5$ and σ can take 21 values (i.e., the Markov chain has 105 states), deviates significantly from the success probability[2] observed in real runs. Figure 6 shows how accurate the predictions of the model can be throughout a run. Note that as one increases the grid resolution, the size of the element containing the optimum, $|\Omega_n|$, decreases. So, it becomes harder and harder to hit such a region (for both the model and the algorithm). This is the reason why in Table 1 the figures for $n = 20$ are bigger than for $n = 40$.

7.2 Bare-Bones PSO

In the experiments with the bare bones PSO we performed 5,000 runs for each setting. Runs lasted 100 generations. In this case we wanted to compare not just the success probability, but the whole state distribution at the end of the runs.

Figures 7–12 compare the distributions obtained in real runs with those predicted by the chain for Ramp and Deceptive and for population sizes $P = 2$, $P = 3$ and $P = 4$ in the case where the domain is divided into just $n = 5$ subdomains. Because the number of states grows very quickly with the resolution,

---

[2] More precisely, the probability of sampling Ω_n.

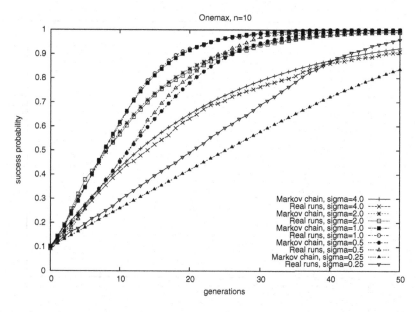

Fig. 2. $(1 + 1)$-ES with fixed σ: comparison between the success probability predicted by the chain and that recorded in real runs for the Ramp function discretised with $n = 10$ plateaus and for different values of σ

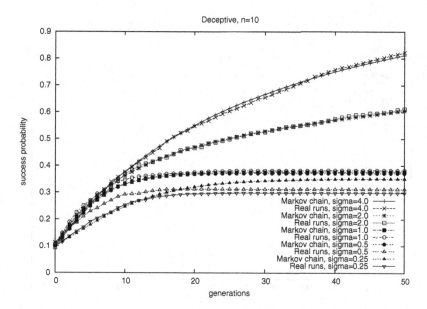

Fig. 3. $(1 + 1)$-ES with fixed σ: comparison between the success probability predicted by the chain and that recorded in real runs for the Deceptive function discretised with $n = 10$ plateaus and for different values of σ

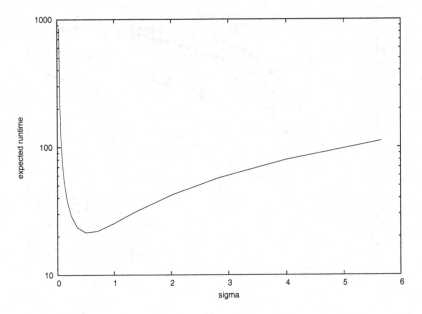

Fig. 4. $(1 + 1)$-ES with fixed σ: expected first hitting time for the domain containing the global optimum for the Ramp function discretised with $n = 40$ plateaus

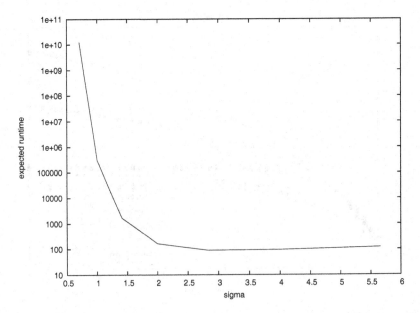

Fig. 5. $(1 + 1)$-ES with fixed σ: expected first hitting time for the domain containing the global optimum for the Deceptive function discretised with $n = 40$ plateaus

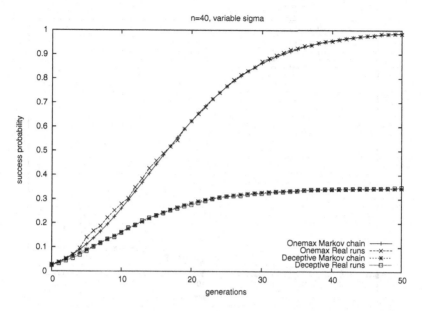

Fig. 6. Comparison between the probability of sampling the optimum domain predicted by the Markov chain for a variable-σ $(1+1)$-ES and empirical data (averages over 1,000 independent runs) on Ramp and Deceptive. Grid with $n = 40$ elements.

n, and the population size, P, and only very few states have non-zero probabilities, it is very hard to obtain a meaningful plot of the full state distribution. So we plot only the 20 states with the largest probabilities. Despite the low resolution used, we obtain an extremely good match between the distributions for all population sizes tested. Also, the success probabilities (the rightmost point in the plots for Ramp, and the one just before the last for Deceptive) match very closely.

Naturally, as with the ES models, increasing the resolution n (see Figure 7) improves fidelity and provides more accurate information on the distribution of the population.

7.3   Real-Valued GA

We performed 5,000 real-valued GA runs for each parameter setting. Runs lasted 100 generations. As with the PSO, we focused on the state distribution at the end of the runs.

Figures 13–18 compare the distributions obtained in real runs with those predicted by the chain for the case where the domain is divided into just $n = 5$ sub-domains and for population sizes $P = 2$, $P = 3$ and $P = 4$. Again, the figures plot the 20 states with the largest probabilities. Despite the low resolution used, we obtain an extremely good match between the distributions for all population sizes tested and the success probabilities match very closely.

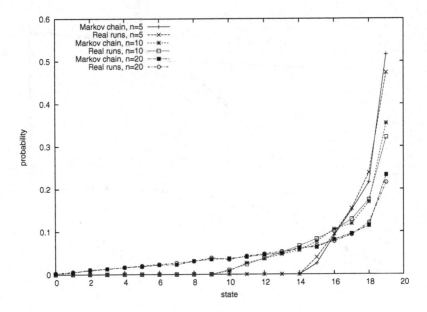

Fig. 7. Comparison between predicted and observed state distributions at generation 100 for a BB-PSO (population size $P = 2$) applied to the Ramp function and for grid resolutions $n = 5$, $n = 10$ and $n = 20$

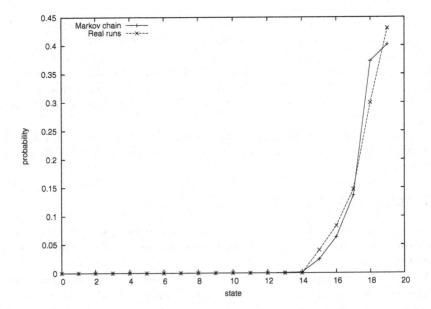

Fig. 8. Comparison between predicted and observed state distributions at generation 100 for a BB-PSO (population size $P = 2$) applied to the Deceptive function and for a grid resolutions $n = 5$

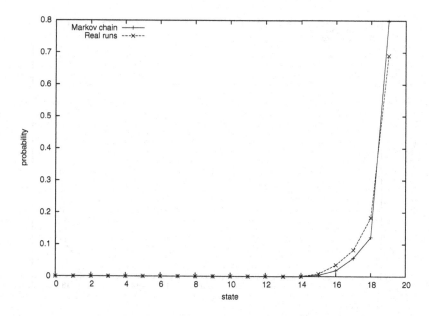

Fig. 9. As in Figure 7 but for $P = 3$ and $n = 5$

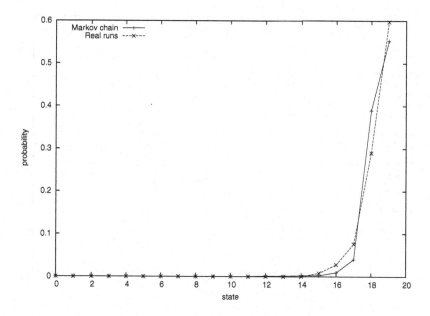

Fig. 10. As in Figure 8 but for $P = 3$

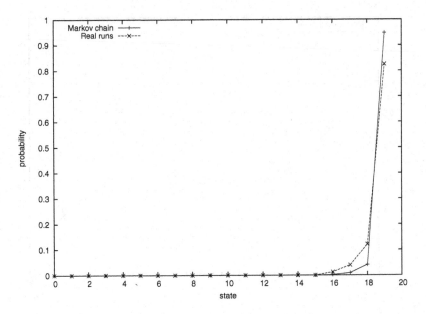

Fig. 11. As in Figure 7 but for $P = 4$ and $n = 5$

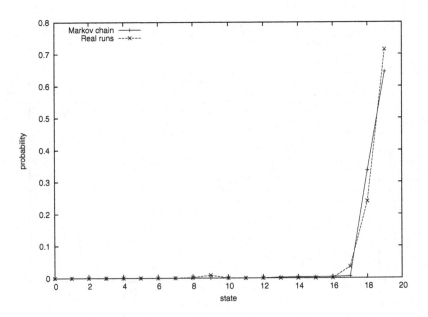

Fig. 12. As in Figure 8 but for $P = 4$

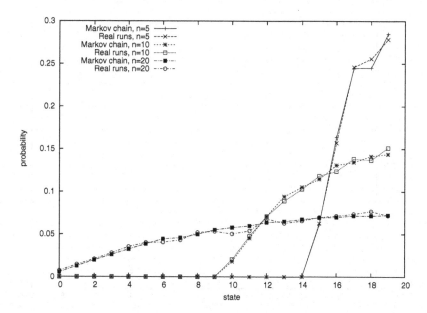

Fig. 13. Comparison between predicted and observed state distributions at generation 100 for a GA (population size $P = 2$) applied to the Ramp function and for grid resolutions $n = 5$, $n = 10$ and $n = 20$

Again, increasing the resolution n (see Figure 13) provides more accurate information on the distribution of the population.

7.4 Two-Dimensional Problems: Sphere and Rastrigin

Very accurate results can also be obtained for higher dimensional and realistic test functions. Figure 19, for example, compares the success probability estimated by the chain and the actual success rate in 100,000 independent runs for the variable-σ $(1 + 1)$-ES used in Section 7.1 on a 2–D sphere function over the interval $[-5, 5)^2$ discretise with a $21 \times 21 = 441$ element grid. Since we allowed 21 different σ's, the total number of states in the Markov chain was 9261. This might appear large, however the transition matrix is very sparse and the chain can be computed and iterated in minutes on an ordinary personal computer.

Results of a similar quality were obtained when we applied the approach to a 2–D Rastrigin function over the interval $[-5, 5)^2$. Because of the complexity of this function (it presents 100 optima in the interval chosen), we used a more sophisticated variable meshing technique. The discretisation proceeded at the 21×21 resolution until an element with high fitness was found. When this happened the element was replaced by a set of smaller ones, effectively locally increasing the resolution to that of a 61×61 grid. This gave a finite element grid of 1329 elements instead of the 441 used for the sphere function. Consequently the total number of states was about three times higher, namely 27909. As shown in Figure 20, chain and experiments are in excellent agreement. Note that the

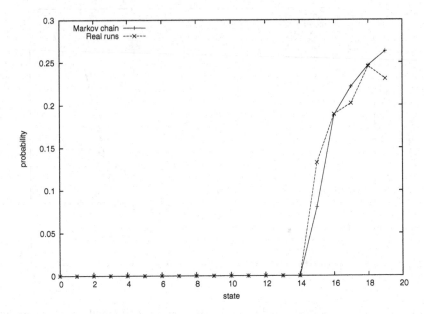

Fig. 14. Comparison between predicted and observed state distributions at generation 100 for a GA (population size $P = 2$) applied to the Deceptive function and for a grid resolutions $n = 5$

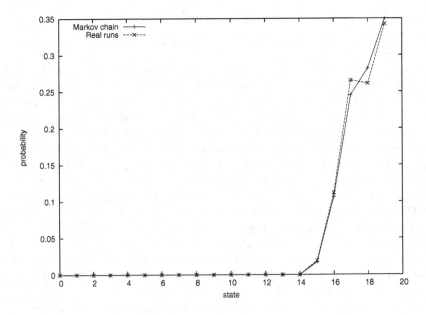

Fig. 15. As in Figure 13 but for $P = 3$ and $n = 5$

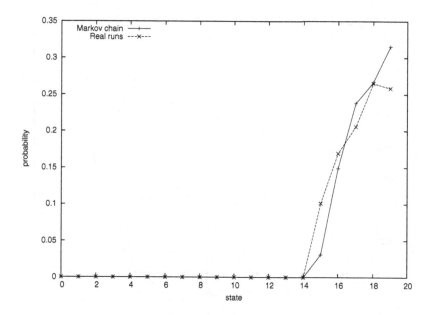

Fig. 16. As in Figure 14 but for $P = 3$

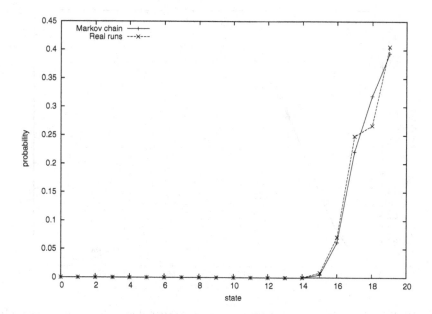

Fig. 17. As in Figure 13 but for $P = 4$ and $n = 5$

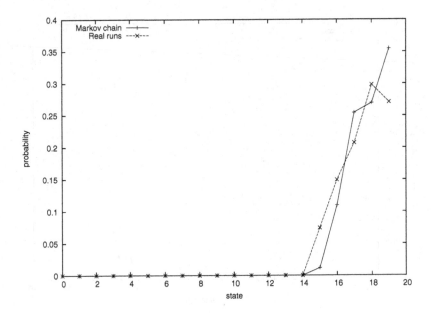

Fig. 18. As in Figure 14 but for $P = 4$

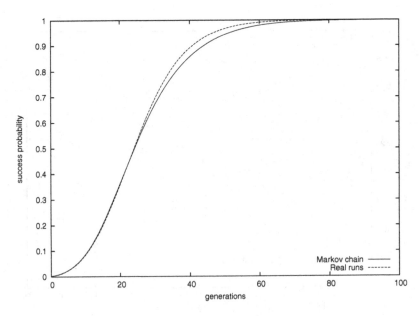

Fig. 19. ES with variable σ: comparison between predicted and observed success probabilities for a 2–D sphere function. Observations are means of 100,000 runs.

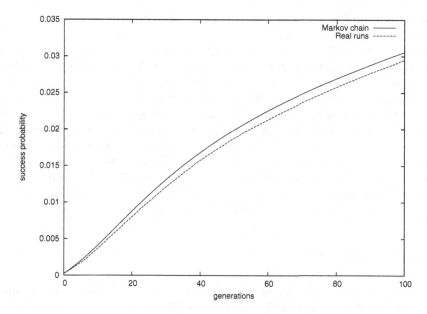

Fig. 20. ES with variable σ: comparison between predicted and observed success probabilities for a 2–D Rastrigin function. Observations are means of 1,000,000 runs.

element containing the optimum is 9 times smaller that for the sphere function. So, to obtain reliable statistics we performed 1,000,000 runs.

8 Conclusions

We have introduced a finite element method to construct discrete Markov chain models for continuous optimisers and we have tested it on two types of evolutionary strategies, on the "bare bones" particle swarm optimiser and on a genetic algorithm with continuous gene values. Whilst the models are approximate, they can be made as accurate as desired by reducing the size of the sub-domains used to quantise the system.

Being Markov chains, the models allow one to compute everything that one needs to estimate about the distribution of states of a search algorithm over any number of generations and for any fitness function. This is a complete characterisation of the behaviour of the search algorithm. For example, in this single framework, in addition to the success probability and the expected runtime, one could calculate the evolution of mean fitness, the population diversity and the size of basins of attraction. We can also compare the behaviour of algorithms by comparing their Markov chains for different problems and compare how different fitness functions influence the behaviour of an algorithm by comparing the corresponding chains.

This is remarkable, but there is of course a price to pay. The price is that, unsurprisingly, like most other models of evolutionary algorithms, the model scales exponentially with the population size, or more generally the size of the memory used by a search algorithm.

In future research we intend to present a deeper analysis and comparison of the Markov chains obtained for different algorithms. We will look at the limiting behaviour of system in the infinite time limit by applying traditional Markov chain analysis techniques. We will also study our models mathematically in the limit of the discretisation resolution going to zero. Finally we want to apply the method to a broader variety of search algorithms, including simulated annealing, traditional particle swarms with velocities, $(\mu+\lambda)$-ESs, and differential evolution.

The method also opens the way to using the mathematical power of Markov chains, specifically existing results on their limiting distribution and rates of convergence, for a far wider range of practical evolutionary algorithms and realistic fitness functions, than has previously been the case.

Acknowledgements

The authors would like to acknowledge support by EPSRC XPS project (GR/T11234/01) and Leverhulme Trust (F/00213/J).

References

1. Babuska, I.: The Finite Element Method for Infinite Domains. I. Mathematics of Computation 26(117), 1–11 (1972)
2. Bäck, T.: Evolutionary Algorithms in Theory and Practice: Evolution Strategies, Evolutionary Programming, Genetic Algorithms. Oxford University Press, New York (1996)
3. Bathe, K.J.: Finite Element Procedures in Engineering Analysis. Prentice-Hall, Englewood Cliffs, New Jersey (1982)
4. Clerc, M.: Stagnation analysis in particle swarm optimisation or what happens when nothing happens. Technical Report Technical Report CSM-460, Department of Computer Science, University of Essex, Edited by Riccardo Poli (August 2006)
5. Clerc, M., Kennedy, J.: The particle swarm - explosion, stability, and convergence in a multidimensional complex space. IEEE Transaction on Evolutionary Computation 6(1), 58–73 (2002)
6. Givoli, D.: Numerical methods for problems in infinite domains. Elsevier, New York (1992)
7. Goldstein, C.: The Finite Element Method with Nonuniform Mesh Sizes for Unbounded Domains. Mathematics of Computation 36(154), 387–404 (1981)
8. Kennedy, J.: The behavior of particles. In: Evolutionary Programming VII: Proceedings of the Seventh Annual Conference on evolutionary programming, pp. 581–589, San Diego, CA (1998)
9. Kennedy, J.: Bare bones particle swarms. In: Proceedings of the IEEE Swarm Intelligence Symposium 2003 (SIS 2003), pp. 80–87, Indianapolis, Indiana (2003)
10. Kennedy, J., Eberhart, R.C., Shi, Y.: Swarm Intelligence. Morgan Kaufmann, San Francisco (2001)

11. Krohling, R.A.: Gaussian particle swarm with jumps. In: Corne, D., Michalewicz, Z., Dorigo, M., Eiben, G., Fogel, D., Fonseca, C., Greenwood, G., Chen, T.K., Raidl, G., Zalzala, A., Lucas, S., Paechter, B., Willies, J., Guervos, J.J.M., Eberbach, E., McKay, B., Channon, A., Tiwari, A., Volkert, L.G., Ashlock, D., Schoenauer, M. (eds.) Proceedings of the 2005 IEEE Congress on Evolutionary Computation, Edinburgh, UK, September 2-5, 2005, vol. 2, pp. 1226–1231. IEEE Press, New York (2005)

12. Nomura, T., Shimohara, K.: An analysis of two-parent recombinations for real-valued chromosomes in an infinite population. Evolutionary Computation 9(3), 283–308 (2001)

13. Ozcan, E., Mohan, C.K.: Particle swarm optimization: surfing the waves. In: Proceedings of the IEEE Congress on evolutionary computation (CEC 1999), Washington, DC (1999)

14. Rudolph, G.: Convergence of evolutionary algorithms in general search spaces. In: International Conference on Evolutionary Computation, pp. 50–54 (1996)

15. Spears, W.M.: The Role of Mutation and Recombination in Evolutionary Algorithms. PhD thesis, George Mason University, Fairfax, Virginia, USA (1998)

16. Storn, R.: Designing digital filters with differential evolution. In: Corne, D., Dorigo, M., Glover, F. (eds.) New Ideas in Optimization, Advanced Topics in Computer Science, ch. 7, pp. 109–125. McGraw-Hill, Maidenhead, Berkshire, England (1999)

17. Thompson, L., Pinsky, P.: Space-time finite element method for structural acoustics in infinite domains. Part 1: formulation, stability and convergence. Computer Methods in Applied Mechanics and Engineering 132(3), 195–227 (1996)

18. Trelea, I.C.: The particle swarm optimization algorithm: convergence analysis and parameter selection. Information Processing Letters 85(6), 317–325 (2003)

19. van den Bergh, F.: An Analysis of Particle Swarm Optimizers. PhD thesis, Department of Computer Science, University of Pretoria, Pretoria, South Africa (November 2001)

Saddles and Barrier in Landscapes of Generalized Search Operators

Christoph Flamm[1], Ivo L. Hofacker[1], Bärbel M.R. Stadler[2],
and Peter F. Stadler[3,1,4]

[1] Department of Theoretical Chemistry University of Vienna,
Währingerstraße 17, A-1090 Wien, Austria
{xtof,ivo,baer,studla}@tbi.univie.ac.at
[2] Max Planck Institute for Mathematics in the Sciences,
Inselstrasse 22-26, D-04103 Leipzig, Germany
[3] Bioinformatics Group, Department of Computer Science, and
Interdisciplinary Center for Bioinformatics (IZBI), University of Leipzig,
Härtelstraße 16-18, D-04107 Leipzig, Germany
{xtof,ivo,baer,studla}@bioinf.uni-leipzig.de
Tel.: ++49 341 97-16691; Fax: ++49 341 97-16709
[4] Santa Fe Institute,
1399 Hyde Park Rd., Santa Fe, NM 87501, USA

Abstract. Barrier trees are a convenient way of representing the structure of complex combinatorial landscapes over graphs. Here we generalize the concept of barrier trees to landscapes defined over general multi-parent search operators based on a suitable notion of topological connectedness that depends explicitly on the search operator. We show that in the case of recombination spaces, path-connectedness coincides with connectedness as defined by the mutation operator alone. In contrast, topological connectedness is more general and depends on the details of the recombination operators as well. Barrier trees can be meaningfully defined for both concepts of connectedness.

1   Introduction

The concept of energy landscapes has proven to be of fundamental relevance in investigations of complex disordered systems, from simple spin glass models to biopolymer folding. Barrier trees [1,2,3,4,5] provide a convenient condensed representation of the discrete landscapes such as the energy landscapes of biopolymer folding and the fitness landscapes of complex combinatorial optimization problems. Barrier trees encapsulate information of mutual reachability of local optima and the energy/fitness barriers that separate them. The concept easily generalizes to PO-set-valued landscapes, which arise naturally in multi-objective optimization [6]. In most studies it has however been restricted to mutation (single parent) as the search operator. In [7,8], barrier trees are used for studying heuristic optimization algorithms including genetic algorithms. In this work the barrier trees are built relative to the Hamming (bit-flip) neighborhood, i.e., without regard to the structure of the underlying search operator.

C.R. Stephens et al. (Eds.): FOGA 2007, LNCS 4436, pp. 194–212, 2007.

In the case of multi-parent search operators it is not obvious how barrier trees should be defined in such a way that the structure of the search spaces that is induced by the search operators is faithfully represented. In a series of papers [9,10,11,12] we have explored how generalized topology can be used to describe the search spaces underlying evolutionary processes with recombination and chemical reaction networks. In the latter, the educts and products of chemical reactions take on the roles of the parents and offsprings, respectively. In this contribution we first demonstrate that search spaces of combinatorial optimization problems inherit a definite topological structure from the collection of search operators that is used by a particular algorithm, such as Simulated Annealing, a Genetic Algorithm, or Genetic Programming. We then show that this topological structure implies a natural concept of connectedness.

The connectedness of subsets in a search space is a property that, intuitively, should have a close relation to properties of reachability or accessibility. Such notions, however, lie at the heart of theories that explain the performance of heuristic optimization procedures on value landscapes. The simplest example are the "cycles" in the theory of Simulated Annealing [13], which in essence can be understood as the connected components of a subset of search space on which the cost function has values better than a given threshold η. We will see that, in conjunction with the cost function, connectedness of subsets then defines a structure of basins and barriers that generalizes the notion of barrier trees from graphs to spaces induced by arbitrary search operators. We finally give a brief example demonstrating that such a type of landscape analysis is indeed computationally feasible at least for certain Genetic Algorithms.

2   Search Operators and Generalized Topology

A (combinatorial) optimization problem is usually specified in terms of a set X of configurations and a cost function $f : X \to \mathbf{R}$, where $\mathbf{R}$ is an ordered set. In the case of multi-objective optimization [14] we have to admit partially ordered value sets $\mathbf{R}$ [6]. A large class of heuristic algorithms, including Simulated Annealing, Genetic Algorithms, Evolutionary Strategies, or Genetic Programming, attempt to find optimal solutions by moving through the set X and evaluating the cost function at different points $x \in X$. This search procedure imposes an implicit mathematical structure on the set X that determines how points or, more generally, subsets are mutually accessible. In a more biologically inspired setting, this *search space* is uniquely determined by the genetic operators at work: mutation, recombination, genome rearrangements, and so on.

2.1   Mutation and Move Sets

In the case of point mutations and constant length sequences, the situation is straightforward. Naturally, sequences that differ by a single mutation are *neighbors* in "sequence space" [15,16]. The sequence space can thus be represented as a graph, also known as Hamming graph or generalized hypercube. The Hamming distance, d_H, counts the number of positions at which two sequences differ.

Move sets are by no means restricted to mutating letters in fixed length string representations. Other examples that are commonly used in an evolutionary optimization context are permutation operators (e.g. for tours of Traveling Salesman Problems) or the exchange operator (e.g. for Graph Bipartitioning), see [17] and the references therein. These moves, which depend on a single "parent", not on a population, naturally define *edges* on the set X of configurations. Therefore, the search space has the structure of a graph. Obviously, this graph is connected if and only if the move set is ergodic.

2.2   Recombination Spaces

The situation becomes more complicated, however, when recombination (crossover) is considered [18]. The analogue of the adjacency relation of the graph is the recombination set $\mathcal{R}(x, y)$, which is defined as the set of all (possible) recombinants of two parents x and y. Recombination sets satisfy at least two axioms:

(X1) $\{x, y\} \in \mathcal{R}(x, y)$,
(X2) $\mathcal{R}(x, y) = \mathcal{R}(y, x)$.

Condition (X1) states that replication may occur without recombination, and (X2) means that the role of the two parents is interchangeable. Often a third condition

(X3) $\mathcal{R}(x, x) = \{x\}$

is assumed, which is, however, not satisfied by models of unequal crossover [19,9]. Functions $\mathcal{R} : X \times X \to \mathcal{P}(X)$ satisfying (X1-X3) were considered recently as *transit functions* [20] and as *P-structures* [21,22].

In the case of strings of fixed lengths n one requires additional properties. We write x_i for the i-th letter in string x. We may assume a different alphabet $\mathcal{A}_i$ for each position i. While for GAs one usually has $\mathcal{A}_i = \{0, 1\}$ for all positions i, one may have a different number alleles for different genes e.g. in a population genetic setting.

(X4) $\mathcal{R}(x, y) \subseteq \text{span}\{x, y\}$, where

$$\text{span}A = \{z \in X | \forall i : \exists x \in A : z_i = x_i\} \tag{1}$$

is the *linear span* of a set A.
(X5) For $x, y \in X$ and $i \neq j$ there is a recombinant $z \in \mathcal{R}(x, y)$ with $z_i = x_i$ and $z_j = y_j$.

The linear span span$A$ correspond to Antonisse's definition of a schema [23]. It also can be interpreted as a "hyperplane" in the Hamming graph with vertex set X and Hamming neighborhood on $X = \prod_{i=1}^{n} \mathcal{A}_i$, see e.g. [24]. We will not pursue the vector spaces aspects of this construction here, however.

It follows directly from equ.(1) that span is idempotent:

$$\text{span}(\text{span}A) = \text{span}A. \tag{2}$$

For string recombination operators, (X4) implies (X3) since span$\{x\} = \{x\}$. Furthermore, we note that for *uniform crossover*, $\mathcal{R}(x,y) = \text{span}\{x,y\}$.

Condition (X5) characterizes *proper recombination operators* in which any two sequence positions can be separated by cross-over. Note that strict 2-point cross-over (i.e., exactly two break-points *within* the strings) is not proper in this sense, since the first and the last sequence position always stay together in the offsprings. The more common definition, which calls for at most break-points, is of course proper.

We note for later usage we collect here a few simple properties of recombination spaces.

Lemma 1. $z \in \mathcal{R}(x,y)$ *implies* span$\{x,z\} \subseteq$ span$\{x,y\}$.

Proof. $u \in \text{span}\{x,z\}$ iff $u_i = x_i$ or $u_i = z_i$ for $i = 1,\ldots,n$ and $z \in \text{span}\{x,y\}$ iff $z_i = x_i$ or $z_i = y_i$. Thus $u_i = x_i$ or $u_i = y_i$, and hence $u \in \text{span}\{x,y\}$.

Lemma 2. *Let* $\mathcal{R}$ *be a proper recombination operator. Then* $\mathcal{R}(x,y) \setminus \{x,y\} \neq \emptyset$ *if and only if* $d_H(x,y) \geq 2$.

Proof. If $d_H(x,y) \geq 2$ then there are two sequence positions $i \neq j$ such that $x_i \neq y_i$ and $x_j \neq y_j$. Since $\mathcal{R}$ separates i and j, there is a recombinant z with $z_i = x_i \neq y_i$ and $z_j = y_j \neq x_j$, i.e., $z \neq x$ and $z \neq y$.

In the case of Genetic algorithms it seems natural to define reachability via the union of those of mutation and recombination.

2.3 Closure Functions

In the most general case we are given a collection $\mathfrak{X}$ of pairs $(A, B[A])$ where $A, B[A] \subseteq X$ and $B[A]$ are interpreted as the offsprings that are generated from A. In the case of a genetic algorithm, for example, $\mathcal{X}$ encodes both mutational offsprings from individual sequences and pair-wise cross-over products. The collection $\mathfrak{X}$ can be extended to a set-valued set-function $c : \mathcal{P}(X) \to \mathcal{P}(X)$ that describes for **each** subset A of X the collection of all possible offsprings, i.e., the set of points in X that are *accessible* from A by a single application of the search operator:

$$c(A) = \bigcup \left\{ B[A'] \middle| A' \subseteq A \wedge A' \neq \emptyset \wedge (A', B[A']) \in \mathfrak{X} \right\} \tag{3}$$

The condition $A' \neq \emptyset$ prohibits "spontaneous creation", i.e., enforces $c(\emptyset) = \emptyset$. In the case of recombination operators on strings, for example, we have

$$c_{\mathcal{R}}(A) = \bigcup_{x,y \in A} \mathcal{R}(x,y). \tag{4}$$

Equ.(4) together with the definition of the linear span implies

$$c_{\mathcal{R}}(A) \subseteq \text{span} A \tag{5}$$

The function c defined in equ. (3) is a *closure function* in the sense of generalized topology. Indeed, following Kuratowski [25], topological spaces are often defined in terms of such a closure function instead of open or closed sets:

(K0) $c(\emptyset) = \emptyset$.
(K1) $A \subseteq B$ implies $c(A) \subseteq c(B)$ (isotonic).
(K2) $A \subseteq c(A)$ (expanding).
(K3) $c(A \cup B) \subseteq c(A) \cup c(B)$ (sub-additive).
(K4) $c(c(A)) = c(A)$ (idempotent).

In our case, however, we can only verify that the first three axioms will hold. (K0) and (K1) follow directly from the construction of c from $\mathfrak{X}$. Such spaces are *isotone*. In addition, axiom (K2) is satisfied if and only if the parental genotypes *can* be transmitted to the next generation. This type of space is known as *neighborhood space*. Axiom (K3) holds for mutation only models, in which the set of offsprings are generated from a parent without regard to the rest of the population. In the finite case, spaces satisfying (K0) through (K3) are exactly the finite graphs. No good argument can be made for idempotency (K4) in our setting. The structure of search spaces thus is strictly weaker than that of topological spaces. The recombination closure space $(X, c_{\mathcal{R}})$, equ.(4) is an example of a finite neighborhood space. Such structure we recently studied as "$\mathfrak{N}$-structures" [26].

If X is finite, we can obtain a an idempotent function $\bar{c}$ by repeated application of c:

$$\bar{c}(A) = c^N(A) = \underbrace{c(c(c(\ldots(c(A)\ldots)))}_{N \text{ times}} \tag{6}$$

for large enough N. (This construction works also in the infinite case, where N is in general an ordinal number.) In the most prominent cases of recombination operators, 1-point crossover and uniform crossover, it is not hard to verify that

$$\bar{c}_{\mathcal{R}}(A) = \text{span} A \tag{7}$$

In [27] it is shown that $\bar{c}_{\mathcal{R}}(A)$ is always a schema in the sense of Antonisse for the "usual" string recombination operators.

The idempotent closure function $\bar{c}$, in contrast to c, gives a rather coarse grained description of the search space. Furthermore, it is known that a meaningful not-trivial topological theory can be constructed without the (K4) axiom. Indeed, Eduard Čech [28] wrote a classical treatise of point set topology based on non-idempotent closure functions. Even more general spaces, lacking also the additivity assumption (K3) were also considered in the literature, see e.g. [29,30,31,32,33].

As in classical topology on can speak about the interior of a set ($I(A) = X \setminus c(X \setminus A)$) and of neighborhoods N of a point x ($N \in \mathcal{N}(x)$ iff $x \in I(N)$) or of a set A ($N \in \mathcal{N}(A)$ iff $A \subseteq I(N)$). Both the interior function I and the neighborhood systems $\mathcal{N}$ of individual points can be used as alternative, equivalent, definitions of the same mathematical structures. We refer to [10,34] for details on this topic our previous papers for details on this topic. In the following discussion we will mostly avoid the use of interior and neighborhood functions.

2.4 Continuity

The notion of continuity lies at the heart of topological theory. Its importance is emphasized by a large number of equivalent definitions, see e.g. [35,36]. Let (X, c) and (Y, c) be two isotone spaces, i.e., spaces satisfying (K0) and (K1). Then $f : X \to Y$ is continuous if one (and hence all) of the following equivalent conditions holds:

(i) $c\left(f^{-1}(B)\right) \subseteq f^{-1}(c\,(B))$ for all $B \in \mathcal{P}(Y)$.
(ii) $f^{-1}(IB) \subseteq If^{-1}(B)$ for all $B \in \mathcal{P}(Y)$.
(iii) $B \in \mathcal{N}(f(x))$ implies $f^{-1}(B) \in \mathcal{N}(x)$ for all $x \in X$.
(iv) $f(c\,(A)) \subseteq c\,(f(A))$ for all $A \in \mathcal{P}(X)$.

3 Connectedness

3.1 Topological Connectedness

Topological connectedness is closely related to separation. Two sets $A, B \in \mathcal{P}(X)$ are *semi-separated* if there are neighborhoods $N' \in \mathcal{N}(A)$ and $N'' \in \mathcal{N}(B)$ such that $A \cap N'' = N' \cap B = \emptyset$. A set $Z \in \mathcal{P}(X)$ is *connected* in a space (X, c) if it is not a disjoint union of nontrivial semi-separated pairs of non-empty sets $A, Z \setminus A$. There have been several attempts to use connectedness as the primitive notion in topological theory [37,38,39].

If (X, c) is isotone (as we shall assume throughout this manuscript) then A and B are semi-separated if and only if $c\,(A) \cap B = A \cap c\,(B) = \emptyset$. Connectedness in isotonic spaces can thus be characterized by the Hausdorff-Lennes condition: A set $Z \in \mathcal{P}(X)$ is *connected* in an isotonic space (X, c) if and only if for each proper subset $A \subseteq Z$ holds

$$[c\,(A) \cap (Z \setminus A)] \cup [c\,(Z \setminus A) \cap A] \neq \emptyset \tag{8}$$

The collection of connected sets satisfies the following three properties in isotonic spaces [10,40]:

(c1) If Z consists of a single point, then Z is connected.
(c2) If Y and Z are connected and $Y \cap Z \neq \emptyset$ then $Y \cup Z$ is connected
(c3) If Z is connected and $Z \subseteq c\,(Z)$, then $c\,(Z)$ is also connected.
(c4) Let I be an arbitrary index set and $x \in X$. Suppose Z_i is connected and $x \in Z_i$ for $i \in I$. Then $W := \bigcup_{i \in I} Z_i$ is connected.

As a short example of the formalism, we give here an elementary proof of property (c4). We first observe the following simple fact: Suppose A and B are semi-separated and $A' \subseteq A$, $B' \subseteq B$; then A' and B' are also semi-separated. Now suppose W as defined above is not connected, i.e., there is a semi-separation $W = W' \dot{\cup} W''$. Assume w.l.o.g. $x \in W'$. Since the Z_i collectively cover W, there is a set Z_v such that $Z_v \cap W'' \neq \emptyset$. Since $x \in Z_v$ we also have $Z_v \cap W' \neq \emptyset$. Since W' and W'' are semi-separated, $Z_v \cap W'$ and $Z_v \cap W''$ are also semiseparated, i.e., $(Z_v \cap W') \dot{\cup} (Z_v \cap W'')$ is a semiseparation of Z_v, and hence Z_v is not connected, a contradiction.

In particular, a neighborhood space is connected if and only if it is not the disjoint union of two closed (open) sets [40, Thm.5.2]. This result generalizes the analogous well-known statement for topological spaces.

Now consider a set $A \subset X$ and a point $x \in A$ and define:

$$A[x] = \bigcup \{ Z \subseteq A \,|\, x \in Z \text{ and } Z \text{ is connected} \} , \tag{9}$$

i.e., $A[x]$ is the union of all connected subsets of A that contain x. By (c4), the set $A[x]$ is itself connected, i.e., $A[x]$ is the unique maximal connected subset of A that contains x. By construction, the collection of subsets $\{A[x]|x \in X\}$ defines a partition of the set A into maximal connected subsets. These sets are called the *connected components* of A. In particular, $X[x]$ is the connected component of x in the entire space.

The relationship of connected components and semi-separations is more complicated than one might guess. The following result for additive spaces matches our intuition:

Lemma 3. *Suppose $A \subseteq X$ has a finite number $k > 1$ of connected components and let $Q \subseteq A$ be a such a connected component. In a Pr's-topology, Q and $A \setminus Q$ are semi-separated.*

Proof. Observe that there is a semiseparation A', $A \setminus A'$ since A is not connected. Suppose $Q \subseteq A'$. Now either $Q = A'$ or A' is not connected. In the latter case there is a set $A'' \subseteq A'$ such that $A'' \dot\cup (A' \setminus A'')$ is a semiseparation and $Q \subseteq A''$. Thus $c(A \setminus A') \cap A'' = \emptyset$, $c(A' \setminus A'') \cap A'' = \emptyset$, and by (K3) $c(A \setminus A'') \cap A'' = c((A \setminus A') \cup (A' \setminus A'')) \cap A'' = (c(A \setminus A') \cup c(A' \setminus A'')) \cap A'' = \emptyset$. Furthermore, $c(A'') \cap (A' \setminus A'') = \emptyset$ and $c(A'') \subseteq c(A')$ implies $c(A'') \cap (A \setminus A') = \emptyset$. $c(A'') \cap (A \setminus A'') = [c(A'') \cap (A' \setminus A'')] \cup [c(A'') \cap (A \setminus A')] = \emptyset$. We conclude that A'' and $A \setminus A''$ are semi-separated. Repeating this argument a finite number of times shows that we can "cut away" parts of A by means of semi-separations until we are left with Q.

This is *not* true in more general neighborhood spaces. Consider the closure space defined by 1-point crossover on strings of length 4 and consider the set $A = \{x = 0000, y = 0011, z = 1100\}$. We have the semi-separations $\{x\}|\{y\}$, $\{x\}|\{z\}$, $\{y\}|\{z\}$, $\{x,y\}|\{z\}$, $\{x,z\}|\{y\}$. Thus the connected components are the isolated points. Nevertheless, $\{y,z\}, \{x\}$ is not a semi-separation since $x \in \mathcal{R}(y,z)$.

In fact, recombination alone does not lead to connected spaces at all.

Theorem 1. *The closure space $(X, c_\mathcal{R})$ is disconnected, i.e., $X[x] = \{x\}$ for all strings x and any string recombination operator $\mathcal{R}$.*

Proof. The search space is $X = \prod_{i=1}^{n} \mathcal{A}_i$, where $\mathcal{A}_i$ is the alphabet (or set of alleles) for sequence position i, and consider an arbitrary point $x^* \in X$. Denote by $X_1 = \{(x_1^*, y_2, \ldots, y_n)|y_i \in \mathcal{A}_i, i \geq 2\}$ the "hyperplane" defined by the first coordinate of x^*. Its complement is $X_1' = X \setminus X_1 = \{(y_1, y_2, \ldots, y_n)|y_1 \in \mathcal{A}_1 \setminus \{x_1^*\}, y_i \in \mathcal{A}_i, i \geq 2\}$. By construction, $X_1 \cap X_1' = \emptyset$, $X_1 \cup X_1' = X$, and $c_\mathcal{R}(X_1) \subseteq \mathrm{span} X_1 = X_1$, $c_\mathcal{R}(X_1') \subseteq \mathrm{span} X_1' = X_1'$. Thus $X_1 \dot\cup X_1'$ is a semiseparation of X. It follows, that the connected component of x^* is confined to X_1, in symbols $X[x^*] \subseteq X_1$.

Now we define $X_2 = \{(x_1^*, x_2^*, y_3, \dots, y_n) | y_i \in \mathcal{A}_i, i \geq 3\}$ and $X_2' = \{(x_1^*, y_2, y_3, \dots, y_n) | y_3 \in \mathcal{A}_2 \setminus \{x_2^*\}, y_i \in \mathcal{A}_i, i \geq 3\}$. As above, we have $X_2 \cap X_2' = \emptyset$, $X_2 \cup X_2' = X_1$, and $c_{\mathcal{R}}(X_2) \subseteq \text{span} X_2 = X_2$ and $c_{\mathcal{R}}(X_2') \subseteq \text{span} X_2' = X_2'$, i.e., $X_2 \dot\cup X_2'$ is a semiseparation of X_1 and hence $X[x^*] \subseteq X_2$.

Repeating for X_k, the "hyperplane" defined by the first k coordinates of x^*, and its complement X_k' w.r.t. X_{k-1}, we obtain $X[x^*] \subseteq X_k$ and finally $X[x^*] \subseteq X_n = \{x^*\}$.

In order to meaningfully study connectedness in the context of Genetic Algorithms, we thus have consider connectedness for the closure operators that are derived from the superposition of mutation and crossover. The following statements may serve as examples.

Lemma 4. *Let (X, c_{GA}) be a closure space derived from point mutations and and a string recombination operator $\mathcal{R}$. Then $\text{span} A$ is connected for all $A \subseteq X$. Furthermore, $\mathcal{R}(x, y)$ is connected if $\mathcal{R}$ is uniform crossover or 1-point crossover.*

Proof. Let $x, y \in \text{span} A$. We can convert x into y by exchanging one character that differs between x and y after the other (say from left to right). The string obtained in each step differs in a single position from the previous one and is again contained in $\text{span} A$. It follows that for every $x, y \in \text{span} A$ there is path in $\text{span} A$ that leads with Hamming distance $d_H = 1$ steps from x to y. Thus $\text{span} A$ is connected in c_{GA} because it is connected w.r.t. to mutation contribution to c_{GA} alone. For uniform crossover, $\mathcal{R}(x, y) = \text{span}\{x, y\}$. In the case of 1-point crossover we consider the recombinants in the order in which they arise by crossover after position k. For $k = 1$, the two offsprings are either identical to the two parents, or differ by letter in the first sequence position from one of the parents, i.e., $d_H(x, x^1) \leq 1$ and $d_H(y, y^1) \leq 1$. The offsprings obtained from crossover at position k and $k + 1$ can be divided into two pairs (x^k, x^{k+1}) and (y^k, y^{k+1}) with $d_H(x^k, x^{k+1}) \leq 1$ and $d_H(y^k, y^{k+1}) \leq 1$. For crossover before the end of the string we obtain $d_H(x^{n-1}, y) \leq 1$ and $d_H(y^{n-1}, x)$. Thus the recombination products of x and y are located on two paths connecting x and y in Hamming distance $d_H = 1$ steps, see also [18].

3.2   Productive Connectedness

In [11] a less stringent definition of connectivity is introduced that is in particular suitable for chemical reaction networks.

We say that $A, B \in \mathcal{P}(X)$ are *productively separated* if for all $Z \subseteq A \cup B$ holds

(1) $c(Z \cap A) \cap B = \emptyset$ and $c(Z \cap B) \cap A = \emptyset$

(2) $c(Z) = c(Z \cap A) \cup c(Z \cap B)$.

If (X, c) is an isotonic space, then A and B are semi-separated if condition (1) holds for all $Z \subseteq A \cup B$.

It is now natural to call a set Z *productively connected* if it cannot be decomposed into two non-empty subsets Z' and $Z'' = Z \setminus Z'$ with $Z' \cap Z'' = \emptyset$

that are productively separated. In general, if Z is connected, then it is also productively connected. In pretopological spaces (and in particular in digraphs), semi-separation and productive separation coincide, hence Z is productively connected if and only if it is connected in this case.

Lemma 5. *Let (X, c_R) be the closure space deriving from a proper recombination operator $\mathcal{R}$. Then a 2-point set $\{x, y\} \subseteq X$ is productively connected if and only if $d_H(x, y) \geq 2$.*

Proof. By (X3), $c_R(x) = \{x\}$ for all $x \in X$. Thus $\{x, y\}$ is productively connected if and only if $c(\{x, y\}) \neq \{x, y\}$. By Lemma 2 this is the case if and only if $d_H(x, y) \geq 2$.

If follows immediately, that any subset $A \subseteq X$ is productively connected for any GA-closure c_{GA} that is derived from point mutation and a proper recombination operator. This implies that the notion of productive connectedness is too weak to be of much use for our purposes. We will therefore not consider productive connectedness in the following.

3.3   Path-Connectedness

Path-connectedness is a widely used notion of connectedness that in general is stronger than topological connectedness. From the topological point of view, a *path* is a *continuous* function $p : [0, 1] \to X$ whose endpoints are $p(0)$ and $p(1)$. (Here the interval $[0, 1]$ is assumed to have the usual topology of real number.) A set A is *path-connected* if for any two points $x, y \in A$, there is a path p with $p(0) = x$ and $p(1) = y$. $A \subset X$ is path-connected if and only if for every pair of points $x, y \in A$ there is a path in A with endpoints x and y. One easily checks that the concatenation of two paths with $p_1(1) = p_2(0)$,

$$p_1 \bullet p_2(t) = \begin{cases} p_1(2t) & t \in [0, 1/2] \\ p_2(2t - 1) & t \in [1/2, 1] \end{cases} \tag{10}$$

is again a path. Conversely, if the restriction of a path p to an interval $[t', t''] \subseteq [0, 1]$ is again a path $p'(t) = p(t(t'' - t') + t')$. In the finite case, paths reduce to simple combinatorial objects, as we shall see in the following two results.

Lemma 6. *Let (X, c) be a neighborhood space and $\{x, y\} \subseteq X$ a 2-point subset. Then the following statements are equivalent:*

1. *$y \in c(\{x\})$ or $x \in c(\{y\})$.*
2. *$\{x, y\}$ is path-connected.*
3. *$\{x, y\}$ is connected.*

Proof. (i⇒ii) Suppose $y \in c(\{x\})$. Then $p : [0, 1] \to \{x, y\}$, $p(t) = x$ for $t \in [0, 1/2)$ and $p(t) = y$ for $t \in [1/2, 1]$ is continuous since $p(c([0, 1/2))) = p([0, 1/2]) = \{x, y\} \subseteq c(p([0, 1/2))) = c(\{x\}) = \{x, y\}$ and $p(c([1/2, 1])) =$

$p([1/2, 1]) = \{y\} \subseteq c(p((1/2, 1])) = c(\{y\})$. Analogously, $p : [0, 1] \to \{x, y\}$, $p(t) = x$ for $t \in [0, 1/2]$ and $p(t) = y$ for $t \in (1/2, 1]$ is continuous if $x \in c(\{y\})$.
(ii⇒iii) The continuous image of a connected set is connected [40, Thm.5.4], thus path-connectedness in general implies connectedness.
(iii⇒i) Since $\{x, y\}$ is connected, $\{x\}, \{y\}$ is not a semiseparation, i.e., $x \in c(\{y\})$ or $y \in c(\{x\})$.

Theorem 2. *Let (X, c) be a finite neighborhood space. The there is a path from x to y in $A \subseteq X$, $x, y \in A$, if and only if there is a sequence of points $(x = x_0, x_1, \ldots, x_{\ell-1}, x_\ell = y)$, $x_i \in A$, such that the two-point sets $\{x_{i-1}, x_i\}$ are connected.*

Proof. Since X is finite, we only need to consider paths p along which the function value $p(t)$ changes a finite number of times. Thus we can decompose $p = p_1 \bullet p_2 \bullet \cdots \bullet p_\ell$ into sub-paths $p_i : [0, 1] \to \{x_{i-1}, x_i\}$ that connect subsequent function values, i.e., p is path if and only if each p_i is continuous. By Lemma 6 such a path with subsequent function values $x_0 = x, x_1, \ldots, x_\ell$ exists if and only if each subset $\{x_{i-1}, x_i\}$ is connected.

With a finite neighborhood space (X, c) we can therefore associate a graph Γ with vertex set X and (directed) edges (x, y) if $y \in c(\{x\})$. If c is additive, than Γ is an equivalent representation of (X, c). This correspondence between finite pretopologies and finite digraphs is discussed e.g. in [41]. In general, we can use the graph Γ to represent path-connectedness in (X, c). It follows directly from theorem 2 that a set $A \subseteq X$ is path-connected (w.r.t. c) if and only if the subgraph of Γ induced by A is connected.

We remark that, for an GA-closure deriving from point mutations and a string recombination operator, span$A$ is path-connected for all A. Furthermore, $\mathcal{R}(x, y)$ is path-connected for 1-point crossover.

Theorem 1 implies that in string-recombination-only closures, there are no connected two-point sets (since the connected components are the individual points). It follows that path-connectedness is determined by the mutation component of the GA.

4   Basins and Barriers

Consider a landscape (X, c, f), where $f : X \to \mathbb{R}$ is an arbitrary function. We define the level-sets

$$X_\eta = \{x \in X | f(x) \le \eta\} \tag{11}$$

Let $\mathbb{P}_{xy}$ be the set of all paths from x to y. We say that x and y are *mutually accessible at level η*, in symbols

$$x \leftarrow\!\!\underline{p\ \eta}\!\!\rightarrow y, \tag{12}$$

if there is path $\mathbf{p} \in \mathbb{P}_{xy}$ such that $f(z) \le \eta$ for all $z \in \mathbf{p}$, respectively. The path-connected components of $X\eta$ are therefore

$$P_\eta[x] = \{y \in V | y \leftarrow\!\!\underline{p\ \eta}\!\!\rightarrow x\} \tag{13}$$

So-called cycles play a central role in the theory of simulated annealing, see e.g. [42,43]. In the landscape setting "cycles" correspond to the connected components of the level sets. In the literature on "disconnectivity graphs", the cycles are usually called "super-basins" [44]. More precisely, the *cycle of* $x \in V$ *at height* η, $C_\eta[x]$, is the *connected component* of the level set $\{y \in V | f(y) \le \eta\}$ that contains x. In finite pretopological spaces, i.e., digraphs, we have $P[x] = C[x]$. In finite neighborhood spaces (and in infinite pretopologies), however, we only have $P[x] \subseteq C[x]$.

For simplicity of the following discussion, let us assume that X is finite and the landscape is non-degenerate (i.e., $f(x) = f(y)$ implies $x = y$). In the case, we may order the points in X by increasing cost, i.e., $f(x^{(i)}) < f(x^{(j)})$ iff $i < j$. Thus $x^{(1)}$ is the unique global minimum. We say that a point x is a *local minimum* if $C_{f(x)}[x] = \{x\}$. In the case of non-degenerate landscapes we would use a more complicated definition. For example, we might say that x is a local optimum if $f(y) = f(x)$ for all $y \in C_{f(x)}[x] : f(y) = f(x)$. The complications of degenerate landscapes are discussed in detail in [5] for the case of landscapes on finite graphs.

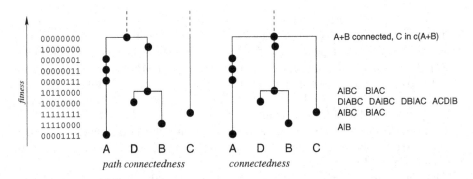

Fig. 1. Example of the lowest part of path-connectedness and connectedness barrier trees for a GA with point mutation and 1-point crossover. The first difference is the simultaneous connection between the connected components A, B, and C via the string 00000000. Just below the fitness level of 00000000, we have A = {00001111, 00000111, 00000011, 00000001}, B = {10000000, 10110000, 10010000, 11110000}, C = {11111111}. Note that A, B, and C are connected by means of point mutations alone. 00000000 connects A and B to a single connected component also via point mutations. Since 11111111 ∈ C is a recombination product of 00001111 ∈ A and 11110000 ∈ B. Since at the level of 00000000, A ∪ B is connected and C = {11111111} is contained in the closure c(A ∪ B), there is no semi-separation in A ∪ B ∪ C, i.e., this set is connected. In the right-most column, we show the complete list of all maximal semi-separations.

In the following we consider the relationship of the (path)connected components of the level sets X_η and $X_{\eta'}$ in some more detail:

Lemma 7. *Let A be a component of X_η and B a component of $X_{\eta'}$ with $\eta < \eta'$. Then either $A \subseteq B$ or $A \cap B = \emptyset$.*

Proof. Consider a point $x \in A$ and let A' be the component of $X_{\eta'}$ that contains x. Since $X_\eta \subseteq X_{\eta'}$, it follows that $A \subseteq A'$. Either $A' = B$, in which case $A \subseteq B$, or $A' \cap B = \emptyset$, in which case $A \cap B = \emptyset$.

Note that this result holds for all notions of connectedness, including topological connectedness, path-connectedness and productive connectedness.

Since f is non-degenerate, we may choose the difference between η and η' so that $X_{\eta'} = X_\eta \cup \{x^*\}$. The "new" point x^* will in general interfere with the connectedness structure of the "old" set X_η. It is important to notice that much of the connectedness structure of of X'_η is inherited from the old level set X_η.

Theorem 3. *Suppose $X_{\eta'}$ and X_η, $\eta' > \eta$ are level sets of a fitness landscape and $X_{\eta'} \setminus X_\eta = \{x^*\}$. Let A' be a connected component of $X_{\eta'}$. Then either $x^* \notin A'$, in which case A' is a connected component of X_η, or $x^* \notin A'$, in which case exactly one of the following three statements is true:*

1. *$A' = \{x^*\}$ is a connected component of $X_{\eta'}$. In this case x^* is a local minimum.*
2. *There is a unique connected component A of X_η such that $A' = A \cup \{x^*\}$.*
3. *There are two or more connected components A_i, $i = 1, 2, \ldots,$ of X_η such that $A' = \bigcup_i A_i \cup \{x^*\}$. In this case we call x^* a saddle point.*

Proof. If $X_{\eta'}$ is connected, there is no component that does not contain x^*. Otherwise there is a semi-separation, say $Q'_1 \dot\cup Q''_1$, of $X_{\eta'}$. Suppose Q''_1 contains x^*. Then $Q'_1 \dot\cup (Q''_1 \setminus \{x\})$ is also a semi-separation of X_η. If $A' \subseteq Q'_1$ it follows immediately that A' is a connected component of Q'_1 and hence also of X_η. We repeat the argument with Q''_1: Unless Q''_1 is connected, there is a semi-separation $Q'_2 \dot\cup Q''_2$ of Q''_1. By the same argument as above, A' lies either in Q'_2 or in $Q''_2 \setminus \{x^*\}$, or it contains x^*. In the first case, A' is a connected component of Q'_2 and hence also of X_η. After a finite number of steps we have either identified A' as a connected component of X_η, or $A' = Q''_k$ is a connected set that contains x^*. In this case $A' = X_{\eta'}[x^*]$

A connected component A_i of X_η is also connected in $X_{\eta'}$. Thus either $A_i \subseteq X_\eta[x^*]$ or $A_i \cap X_\eta[x^*] = \emptyset$. It follows that $X_\eta[x^*] = \bigcup_{i \in I} A_i \cup \{x^*\}$ for a suitable finite index set I. The three cases in the statement of theorem correspond to $|I| = 0$, $|I| = 1$, and $|I| \geq 2$, respectively.

Note that this result is rather trivial in finite pretopologies (i.e., graphs). In this case the connected component of x^* in X'_η is the union of all connected sets $A_i \cup \{x^*\}$, while lemma 3 guarantees that the remaining connected components of X_η are also connected components of X'_η. Unfortunately, this simple construction does not work in non-additive spaces.

Algorithmically, it seems to be useful to keep track not only of the connected components but also of the semi-separations between them. Since semi-separations are inherited by subsets, we can, conversely argue, that a semi-separation of $U \cup \{x\}$ is either of the form $\{x\}|U$ or there is semi-separation $U = U'|U''$ such that $U' \cup \{x\}|U''$ or $U'|U'' \cup \{x\}$ is a semiseparation. Here U, U' and U'' are unions of connected components.

Algorithm 1. Barrier Trees with arbitrary search operators

1: **procedure** BARRIERS
2:     $\mathcal{L}^{act} \leftarrow \text{MERGE}(\mathcal{L}, x)$
3:     **if** $\mathcal{L}^{act} = (\{x\})$ **then**
4:         x is a local minimum
5:     **else if** $\mathcal{L}^{act} = (\{x\} \cup A)$ **then**
6:         x belongs to basin A
7:     **else**
8:         x is a saddle point merge all $A \in \mathcal{L}^{act} \backslash \{x\}$

In practice, one first checks whether the new point $\{x\}$ can be connected to one or more connected components via mutation. If so, there is no semi-separation between $\{x\}$ and these components. For the remaining candidates, it is sufficient to consider recombination. A candidate semi-separation of the form $U' \cup \{x\}|U''$ can be ruled out if there is either $\mathcal{R}(x, z) \cap U'' \neq \emptyset$ for some $z \in U'$ or if there is $y, z \in U''$ such that $x \in \mathcal{R}(y, z)$. Note that, if $U' \cup \{x\}|U''$ can be ruled out as a semi-separation, then $V' \cup \{x\}|V''$ with $U' \subseteq V'$ and $U'' \subseteq V''$ is also not a semi-separation. Thus it is in particular sufficient to compute for every connected component A, whether x is a recombinant of A, which other connected components contain recombinants of x and members of A, and for every pair of components A' and A'', whether x is recombinant of parents in A' and A'', respectively. More precisely, $U' \cup \{x\}|U''$ is a semiseparation, if and only if (1) for all connected components $A \subseteq U'$ and $B \subseteq U''$, $A \cup \{x\}|B$ is a semiseparation and (2) $\{x\}|B' \cup B''$ is semiseparation for all $B', B'' \subseteq U''$. Two connected components of W belong to the same connected component of $W \cup \{x\}$ if and only if there is not semi-separation left that separates them.

Algorithm 1 summarizes the basic logic of computing barrier trees. In algorithm 2 we outline the steps that are necessary to update the collection $\mathcal{L}$ of connected components when a single point x is added for the case of mutation/recombination operators. Our approach relies on updating the list $\mathcal{S}$ of maximal semiseparations. These steps are independent of the details of closure function. In the case of recombination we use the $R(A, B \rightarrow C)$ to store the information whether there are parents $a \in A$ and $b \in B$ that give rise to an offspring in connected component C. If one were to consider search operators that construct offsprings from more than two parents, these data-structures would have to be modified accordingly. In Algorithm 3 we collect simplified presentations of the higher-level procedure utilized in Algorithm 2.

In practice, we use a trie data structure to store the connected components s. This allows a more efficient check of $R(A, C \rightarrow \{x\})$. If the sum of the longest prefix of x in A and the longest suffix of x in C has at least the length of x, then x can be produced via recombination from members of the connected components A and C.

Fig. 2 shows barrier trees for landscapes of quadratic spin glasses with randomly generated interaction coefficients. At least for small instances, here 16 bits, examples of landscapes for which recombination changes the barrier structure are rare.

Algorithm 2. Update Connected Components

    ▷ $\mathcal{L}$ List of connected components
    ▷ $\mathcal{S}$ List maximal semiseparations
    ▷ R 3-dimensional array $R(A, B \to C)$ $\forall A, B, C \in \mathcal{L}$
    ▷ $\mathcal{S}(x)$ point mutation neighbors of x
    ▷ $\mathcal{R}(x, y)$ crossover neighbors of x and y

1: **procedure** MERGE
2:      initialize $\mathcal{L}, \mathcal{S}$ and R

    ▷ Update$\mathcal{R}$
3:      **for all** $A, C \in \mathcal{L}$ **do**
4:          $R(A, \{x\} \to C) = 1$ if $\exists a \in A, c \in C : c \in \mathcal{R}(a, x) \vee c \in \mathcal{M}(x)$
5:          $R(\{x\}, A \to C) = R(A, \{x\} \to C)$
6:          $R(A, C \to \{x\}) = 1$ if $\exists a \in A, c \in C : x \in \mathcal{R}(a, c) \vee x \in \mathcal{M}(a) \cup \mathcal{M}(b)$

    ▷ New list of semiseparations
7:      $\mathcal{S}' \leftarrow$ MAXSEMISEPARASTIONS$(\{x\}|\mathcal{L})$
8:      **for all** $(U'|U'') \in \mathcal{S}$ **do**
9:          $\mathcal{S}' \leftarrow \mathcal{S}' \cup$ MAXSEMISEPARASTIONS$(\{x\} \cup U'|U'')$
10:        $\mathcal{S}' \leftarrow \mathcal{S}' \cup$ MAXSEMISEPARASTIONS$(U'|\{x\} \cup U'')$
11:        Remove duplicates and non maximal elements from $\mathcal{S}'$

    ▷ Update $\mathcal{L}$
12:     $\mathcal{L}^{act} \leftarrow \mathcal{L}$
13:     **while** $\exists U'|U'' \in \mathcal{L}^{act}$ with $\mathcal{L}^{act} \cap (U' \cup U'') \wedge U' \cap \mathcal{L}^{act} \neq \emptyset \wedge U'' \cap \mathcal{L}^{act} \neq \emptyset$ **do**
14:        **if** $\{x\} \in U'$ **then**
15:           $\mathcal{L}^{act} \leftarrow U'$
16:        **else**
17:           $\mathcal{L}^{act} \leftarrow U''$
18:     $X \bigcup\limits_{A \in \mathcal{L}^{act}} A$

    ▷ Update R
19:     $R(X, B \to C) \leftarrow \bigvee\limits_{A \in \mathcal{L}^{act}} R(A, B \to C)$
20:     $R(A, B \to X) \leftarrow \bigvee\limits_{C \in \mathcal{L}^{act}} R(A, B \to C)$
21:     **for all** $A \in \mathcal{L}^{act}$ **do**
22:        **for all** (do$B, C \in \mathcal{L}$)
23:           **remove** $R(A, B \to C), R(B, A \to C), R(B, C \to A)$

    ▷ Update $\mathcal{S}$
24:     **for all** $U'|U'' \in \mathcal{S}'$ **do**
25:        $\mathcal{S} \leftarrow \mathcal{S} \cup U'|U''$ **unless** $\exists A \in U' \cap \mathcal{L}^{act} \wedge \exists B \in U'' \cap \mathcal{L}^{act}$
26:     $\mathcal{L} \leftarrow (\mathcal{L} \backslash \mathcal{L}^{act}) \cup X$
     return $\mathcal{L}^{act}$

Further computational studies will be necessary see if this finding is related to the fact that highly correlated landscapes are usually efficiently searchable by means of hill-climbing along.

Algorithm 3. Update Connected Components

1: **procedure** REMOVESUBSETS
2:     **for all** $U'|U''$ **do**
3:         **if** $\exists V', V'' \in \mathcal{S}'$ with $V' \subset U' \wedge V'' \subset U''$ **then**
4:             remove $V'|V''$ from $\mathcal{S}'$
5:         **if** $U' \subset V' \wedge U'' \subset V''$ **then**
6:             ignore $U'|U''$

7: **procedure** ISSEMISEPARASTION
    $U'|U''$ is a semiseparation if
    $\forall A, B \in U' \wedge C \in U'' : R(A, B \rightarrow C) = 0$
    and $\forall A, B \in U'' \wedge C \in U' : R(A, B \rightarrow C) = 0$.

8: **function** MAXSEMISEPARASTIONS$(U'|U'')$
9:     **if** $U'|U''$ is a semiseparation **then**
10:        return$(U'|U'')$
11:    **else**
12:        subsequently delete each element from U' and U'' and call
13:        MAXSEMISEPARASTIONS if this candidate is not yet checked.

---

5 Discussion

In this contribution we have extended the notion of barrier trees to search spaces of arbitrary structure. Two complications arise beyond the realm of finite graphs: (1) There are several natural notions of connectedness, each of which appears as **the** most natural in different applications. For example, productive connectedness was introduced to properly describe chemical reaction networks, while topological connectedness appears as the natural framework to study Genetic Algorithms. (2) The non-additivity of the close function in non-graphical search spaces poses substantial algorithmic challenges in actually computing barrier trees.

In this contribution we have briefly described a prototypical approach that is, however, not practical for large problems since its runtime and memory requirements are quadratic in the size of the (part of the) search space under investigation. More efficient algorithms, or at least efficient heuristics to check connectedness will be required before this approach can be applied to interestingly large optimization problems.

Interestingly, path-connectedness, which is more stringent than topological connectedness, turns out to be equivalent to "connectedness via mutations only" in the context of genetic algorithms. This results provides an *a posteriori* justification of the approach by Halam and Prügel-Bennet to study the dynamics of GAs by mapping the populations onto a barrier tree based on Hamming-neighborhoods. It also suggests to use a comparison of path-connectedness and topological connectedness trees to study the effects of crossover on a given landscape. By construction, the topological connectedness barrier tree is a homeomorphic image of the path-connectedness tree (since connected components are

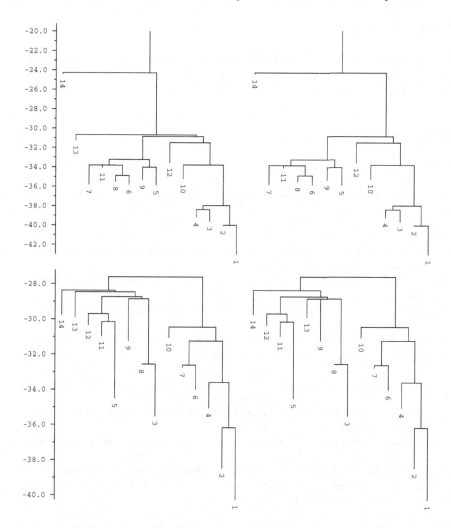

Fig. 2. Two Examples (rows) of barrier trees for quadratic spin glasses of size 16 without and with 1-point recombination (columns). In the first example (upper row) local minimum **13** is lost in the case of recombination because it can be produced from members of local minimum **1**. In the second example, local minimum **13** can be connected to local minimum **3** at a lower energy level in the case of recombination. In both cases the remaining part of the landscape is unaffected.

unions of path-connected components). It follows that the effect of recombination appears as the collapsing of nodes and consequently as the reduction of of the number of local minima and their separating barrier heights. Intuitively, a recombination has big impact if many nodes are collapsed when going from path-connectedness to topological connectedness; conversely, recombination is not helpful when the two trees are (almost) the same.

Acknowledgments

This work was supported in part by the European Union FP-6 as part of the EMBIO project (http://www-embio.ch.cam.ac.uk/) and by the DFG Bioinformatics Initiative (BIZ-6/1-2).

References

1. Klotz, T., Kobe, S.: Valley Structures in the phase space of a finite 3D Ising spin glass with $\pm i$ interactions. J. Phys. A: Math. Gen. 27, L95–L100 (1994)
2. Garstecki, P., Hoang, T.X., Cieplak, M.: Energy landscapes, supergraphs, and folding funnels in spin systems. Phys. Rev. E 60, 3219–3226 (1999)
3. Doye, J.P., Miller, M.A., Welsh, D.J.: Evolution of the potential energy surface with size for Lennard-Jones clusters. J. Chem. Phys. 111, 8417–8429 (1999)
4. Flamm, C., Fontana, W., Hofacker, I., Schuster, P.: RNA folding kinetics at elementary step resolution. RNA 6, 325–338 (2000)
5. Flamm, C., Hofacker, I.L., Stadler, P.F., Wolfinger, M.T.: Barrier trees of degenerate landscapes. Z. Phys. Chem. 216, 155–173 (2002)
6. Stadler, P.F., Flamm, C.: Barrier trees on poset-valued landscapes. Genetic Prog. Evolv. Mach. 4, 7–20 (2003)
7. Hallam, J., Prügel-Bennett, A.: Barrier trees for search analysis. In: Cantú-Paz, E., Foster, J.A., Deb, K., Davis, L., Roy, R., O'Reilly, U.-M., Beyer, H.-G., Kendall, G., Wilson, S.W., Harman, M., Wegener, J., Dasgupta, D., Potter, M.A., Schultz, A., Dowsland, K.A., Jonoska, N., Miller, J., Standish, R.K. (eds.) GECCO 2003. LNCS, vol. 2724, pp. 1586–1587. Springer, Heidelberg (2003)
8. Hallam, J., Prügel-Bennett, A.: Large barrier trees for studying search. IEEE Trans. Evol. Comput. 9, 385–397 (2005)
9. Stadler, B.M.R., Stadler, P.F., Shpak, M., Wagner, G.P.: Recombination spaces, metrics, and pretopologies. Z. Phys. Chem. 216, 217–234 (2002)
10. Stadler, B.M.R., Stadler, P.F.: Generalized topological spaces in evolutionary theory and combinatorial chemistry (Proceedings MCC 2001, Dubrovnik). J.Chem. Inf. Comput. Sci. 42, 577–585 (2002)
11. Benkö, G., Centler, F., Dittrich, P., Flamm, C., Stadler, B.M.R., Stadler, P.F.: A topological approach to chemical organizations. Alife (2006) (submitted)
12. Stadler, P.F., Stadler, B.M.R.: The genotype-phenotype map. Biological Theory 3, 268–279 (2006)
13. Azencott, R.: Simulated Annealing. John Wiley & Sons, New York (1992)
14. Deb, K.: Multi-Objective Optimization using Evolutionary Algorithms. Wiley, Chichester, NY (2001)
15. Maynard-Smith, J.: Natural selection and the concept of a protein space. Nature 225, 563–564 (1970)
16. Eigen, M., Schuster, P.: The Hypercycle. Springer, Heidelberg (1979)
17. Reidys, C.M., Stadler, P.F.: Combinatorial landscapes. SIAM Review 44, 3–54 (2002)
18. Gitchoff, P., Wagner, G.P.: Recombination induced hypergraphs: a new approach to mutation-recombination isomorphism. Complexity 2, 37–43 (1996)

19. Shpak, M., Wagner, G.P.: Asymmetry of configuration space induced by unequal crossover: implications for a mathematical theory of evolutionary innovation. Artificial Life 6, 25–43 (2000)
20. Changat, M., Klavžar, S., Mulder, H.M.: The all-path transit function of a graph. Czech. Math. J. 51, 439–448 (2001)
21. Stadler, P.F., Wagner, G.P.: The algebraic theory of recombination spaces. Evol. Comp. 5, 241–275 (1998)
22. Stadler, P.F., Seitz, R., Wagner, G.P.: Evolvability of complex characters: Population dependent Fourier decomposition of fitness landscapes over recombination spaces. Bull. Math. Biol. 62, 399–428 (2000)
23. Antonisse, J.: A new interpretation of schema notation the overturns the binary encoding constraint. In: Proceedings of the Third International Conference of Genetic Algorithms, pp. 86–97. Morgan Kaufmann, San Francisco (1989)
24. Imrich, W., Klavžar, S.: Product Graphs: Structure and Recognition. Wiley, New York (2000)
25. Kuratowski, C.: Sur la notion de limite topologique d'ensembles. Ann. Soc. Polon. Math. 21, 219–225 (1949)
26. Imrich, W., Stadler, P.F.: A prime factor theorem for a generalized direct product. Discussiones Math. Graph Th. 26, 135–140 (2006)
27. Mitavskiy, B.: Crossover invariant subsets of the search space for evolutionary algorithms. Evol. Comp. 12, 19–46 (2004)
28. Čech, E.: Topological Spaces. Wiley, London (1966)
29. Hammer, P.C.: General topoloy, symmetry, and convexity. Trans. Wisconsin Acad. Sci. Arts, Letters 44, 221–255 (1955)
30. Hammer, P.C.: Extended topology: Set-valued set functions. Nieuw Arch. Wisk. III 10, 55–77 (1962)
31. Day, M.M.: Convergence, closure, and neighborhoods. Duke Math. J. 11, 181–199 (1944)
32. Brissaud, M.M.: Les espaces prétopologiques. C. R. Acad. Sc. Paris Ser. A 280, 705–708 (1975)
33. Gniłka, S.: On extended topologies. I: Closure operators. Ann. Soc. Math. Pol. Ser. I, Commentat. Math. 34, 81–94 (1994)
34. Stadler, B.M.R., Stadler, P.F.: The topology of evolutionary biology. In: Ciobanu (ed.) Modeling in Molecular Biology. Natural Computing Series, pp. 267–286. Springer, Heidelberg (2004)
35. Hammer, P.C.: Extended topology: Continuity I. Portug. Math. 25, 77–93 (1964)
36. Gniłka, S.: On continuity in extended topologies. Ann. Soc. Math. Pol. Ser. I, Commentat. Math. 37, 99–108 (1997)
37. Wallace, A.D.: Separation spaces. Ann. Math, 687–697 (1941)
38. Hammer, P.C.: Extended topology: Connected sets and Wallace separations. Portug. Math. 22, 77–93 (1963)
39. Harris, J.M.: Continuity and separation for point-wise symmetric isotonic closure functions. Technical Report 0507230, arXiv:math.GN (2005)
40. Habil, E.D., Elzenati, K.A.: Connectedness in isotonic spaces. Turk. J. Math 30, 247–262 (2006)
41. Stadler, B.M.R., Stadler, P.F., Wagner, G., Fontana, W.: The topology of the possible: Formal spaces underlying patterns of evolutionary change. J. Theor. Biol. 213, 241–274 (2001)

42. Catoni, O.: Rough large deviation estimates for simulated annealing: Application to exponential schedules. Ann. Probab. 20, 1109–1146 (1992)
43. Catoni, O.: Simulated annealing algorithms and Markov chains with rate transitions. In: Azema, J., Emery, M., Ledoux, M., Yor, M. (eds.) Seminaire de Probabilites XXXIII. Lecture Notes in Mathematics, vol. 709, pp. 69–119. Springer, Heidelberg (1999)
44. Becker, O.M., Karplus, M.: The topology of multidimensional potential energy surfaces: Theory and application to peptide structure and kinetics. J. Chem. Phys. 106, 1495–1517 (1997)

Author Index

Beyer, Hans-Georg   70
Borenstein, Yossi   123
Burjorjee, Keki   35

Cervantes, Jorge   15
Clerc, Maurice   165

Flamm, Christoph   194

Galván-López, Edgar   138
Gedeon, Tomáš   97

Hayes, Christina   97
Hofacker, Ivo L.   194

Jansen, Thomas   54

Langdon, William B.   165

Meyer-Nieberg, Silja   70
Moraglio, Alberto   1

Poli, Riccardo   1, 123, 138, 165

Rowe, Jonathan E.   110

Stadler, Bärbel M.R.   194
Stadler, Peter F.   194
Stephens, Christopher R.   15, 165
Swanson, Richard   97

Vose, Michael D.   110

Wright, Alden H.   110

author index

Lecture Notes in Computer Science

For information about Vols. 1–4489

please contact your bookseller or Springer

Vol. 4600: H. Comon-Lundh, C. Kirchner, H. Kirchner, Rewriting, Computation and Proof. XVI, 273 pages. 2007.

Vol. 4595: D. Bošnački, S. Edelkamp (Eds.), Model Checking Software. X, 285 pages. 2007.

Vol. 4592: Z. Kedad, N. Lammari, E. Métais, F. Meziane, Y. Rezgui (Eds.), Natural Language Processing and Information Systems. XIV, 442 pages. 2007.

Vol. 4591: J. Davies, J. Gibbons (Eds.), Integrated Formal Methods. IX, 660 pages. 2007.

Vol. 4590: W. Damm, H. Hermanns (Eds.), Computer Aided Verification. XV, 562 pages. 2007.

Vol. 4589: J. Münch, P. Abrahamsson (Eds.), Product-Focused Software Process Improvement. XII, 414 pages. 2007.

Vol. 4588: T. Harju, J. Karhumäki, A. Lepistö (Eds.), Developments in Language Theory. XI, 423 pages. 2007.

Vol. 4587: R. Cooper, J. Kennedy (Eds.), Data Management. XIII, 259 pages. 2007.

Vol. 4586: J. Pieprzyk, H. Ghodosi, E. Dawson (Eds.), Information Security and Privacy. XIV, 476 pages. 2007.

Vol. 4584: N. Karssemeijer, B. Lelieveldt (Eds.), Information Processing in Medical Imaging. XX, 777 pages. 2007.

Vol. 4583: S.R. Della Rocca (Ed.), Typed Lambda Calculi and Applications. X, 397 pages. 2007.

Vol. 4582: J. Lopez, P. Samarati, J.L. Ferrer (Eds.), Public Key Infrastructure. XI, 375 pages. 2007.

Vol. 4581: A. Petrenko, M. Veanes, J. Tretmans, W. Grieskamp (Eds.), Testing of Software and Communicating Systems. XII, 379 pages. 2007.

Vol. 4578: F. Masulli, S. Mitra, G. Pasi (Eds.), Fuzzy Logic and Applications. XVIII, 693 pages. 2007. (Sublibrary LNAI).

Vol. 4577: N. Sebe, Y. Liu, Y. Zhuang (Eds.), Multimedia Content Analysis and Mining. XIII, 513 pages. 2007.

Vol. 4576: D. Leivant, R. de Queiroz (Eds.), Logic, Language, Information, and Computation. X, 363 pages. 2007.

Vol. 4574: J. Derrick, J. Vain (Eds.), Formal Techniques for Networked and Distributed Systems – FORTE 2007. XI, 375 pages. 2007.

Vol. 4573: M. Kauers, M. Kerber, R. Miner, W. Windsteiger (Eds.), Towards Mechanized Mathematical Assistants. XIII, 407 pages. 2007. (Sublibrary LNAI).

Vol. 4572: F. Stajano, C. Meadows, S. Capkun, T. Moore (Eds.), Security and Privacy in Ad-hoc and Sensor Networks. X, 247 pages. 2007.

Vol. 4570: H.G. Okuno, M. Ali (Eds.), New Trends in Applied Artificial Intelligence. XXI, 1194 pages. 2007. (Sublibrary LNAI).

Vol. 4569: A. Butz, B. Fisher, A. Krüger, P. Olivier, S. Owada (Eds.), Smart Graphics. IX, 237 pages. 2007.

Vol. 4566: M.J Dainoff (Ed.), Ergonomics and Health Aspects of Work with Computers. XVIII, 390 pages. 2007.

Vol. 4565: D.D. Schmorrow, L.M. Reeves (Eds.), Foundations of Augmented Cognition. XIX, 450 pages. 2007. (Sublibrary LNAI).

Vol. 4564: D. Schuler (Ed.), Online Communities and Social Computing. XVII, 520 pages. 2007.

Vol. 4563: R. Shumaker (Ed.), Virtual Reality. XXII, 762 pages. 2007.

Vol. 4561: V.G. Duffy (Ed.), Digital Human Modeling. XXIII, 1068 pages. 2007.

Vol. 4560: N. Aykin (Ed.), Usability and Internationalization, Part II. XVIII, 576 pages. 2007.

Vol. 4559: N. Aykin (Ed.), Usability and Internationalization, Part I. XVIII, 661 pages. 2007.

Vol. 4549: J. Aspnes, C. Scheideler, A. Arora, S. Madden (Eds.), Distributed Computing in Sensor Systems. XIII, 417 pages. 2007.

Vol. 4548: N. Olivetti (Ed.), Automated Reasoning with Analytic Tableaux and Related Methods. X, 245 pages. 2007. (Sublibrary LNAI).

Vol. 4547: C. Carlet, B. Sunar (Eds.), Arithmetic of Finite Fields. XI, 355 pages. 2007.

Vol. 4546: J. Kleijn, A. Yakovlev (Eds.), Petri Nets and Other Models of Concurrency – ICATPN 2007. XI, 515 pages. 2007.

Vol. 4545: H. Anai, K. Horimoto, T. Kutsia (Eds.), Algebraic Biology. XIII, 379 pages. 2007.

Vol. 4544: S. Cohen-Boulakia, V. Tannen (Eds.), Data Integration in the Life Sciences. XI, 282 pages. 2007. (Sublibrary LNBI).

Vol. 4543: A.K. Bandara, M. Burgess (Eds.), Inter-Domain Management. XII, 237 pages. 2007.

Vol. 4542: P. Sawyer, B. Paech, P. Heymans (Eds.), Requirements Engineering: Foundation for Software Quality. IX, 384 pages. 2007.

Vol. 4541: T. Okadome, T. Yamazaki, M. Makhtari (Eds.), Pervasive Computing for Quality of Life Enhancement. IX, 248 pages. 2007.

Vol. 4539: N.H. Bshouty, C. Gentile (Eds.), Learning Theory. XII, 634 pages. 2007. (Sublibrary LNAI).

Vol. 4538: F. Escolano, M. Vento (Eds.), Graph-Based Representations in Pattern Recognition. XII, 416 pages. 2007.

Vol. 4537: K.C.-C. Chang, W. Wang, L. Chen, C.A. Ellis, C.-H. Hsu, A.C. Tsoi, H. Wang (Eds.), Advances in Web and Network Technologies, and Information Management. XXIII, 707 pages. 2007.

Vol. 4536: G. Concas, E. Damiani, M. Scotto, G. Succi (Eds.), Agile Processes in Software Engineering and Extreme Programming. XV, 276 pages. 2007.

Vol. 4534: I. Tomkos, F. Neri, J. Solé Pareta, X. Masip Bruin, S. Sánchez Lopez (Eds.), Optical Network Design and Modeling. XI, 460 pages. 2007.

Vol. 4531: J. Indulska, K. Raymond (Eds.), Distributed Applications and Interoperable Systems. XI, 337 pages. 2007.

Vol. 4530: D.H. Akehurst, R. Vogel, R.F. Paige (Eds.), Model Driven Architecture- Foundations and Applications. X, 219 pages. 2007.

Vol. 4529: P. Melin, O. Castillo, L.T. Aguilar, J. Kacprzyk, W. Pedrycz (Eds.), Foundations of Fuzzy Logic and Soft Computing. XIX, 830 pages. 2007. (Sublibrary LNAI).

Vol. 4528: J. Mira, J.R. Álvarez (Eds.), Nature Inspired Problem-Solving Methods in Knowledge Engineering, Part II. XXII, 650 pages. 2007.

Vol. 4527: J. Mira, J.R. Álvarez (Eds.), Bio-inspired Modeling of Cognitive Tasks, Part I. XXII, 630 pages. 2007.

Vol. 4526: M. Malek, M. Reitenspieß, A. van Moorsel (Eds.), Service Availability. X, 155 pages. 2007.

Vol. 4525: C. Demetrescu (Ed.), Experimental Algorithms. XIII, 448 pages. 2007.

Vol. 4524: M. Marchiori, J.Z. Pan, C.d.S. Marie (Eds.), Web Reasoning and Rule Systems. XI, 382 pages. 2007.

Vol. 4523: Y.-H. Lee, H.-N. Kim, J. Kim, Y. Park, L.T. Yang, S.W. Kim (Eds.), Embedded Software and Systems. XIX, 829 pages. 2007.

Vol. 4522: B.K. Ersbøll, K.S. Pedersen (Eds.), Image Analysis. XVIII, 989 pages. 2007.

Vol. 4521: J. Katz, M. Yung (Eds.), Applied Cryptography and Network Security. XIII, 498 pages. 2007.

Vol. 4519: E. Franconi, M. Kifer, W. May (Eds.), The Semantic Web: Research and Applications. XVIII, 830 pages. 2007.

Vol. 4517: F. Boavida, E. Monteiro, S. Mascolo, Y. Koucheryavy (Eds.), Wired/Wireless Internet Communications. XIV, 382 pages. 2007.

Vol. 4516: L. Mason, T. Drwiega, J. Yan (Eds.), Managing Traffic Performance in Converged Networks. XXIII, 1191 pages. 2007.

Vol. 4515: M. Naor (Ed.), Advances in Cryptology - EUROCRYPT 2007. XIII, 591 pages. 2007.

Vol. 4514: S.N. Artemov, A. Nerode (Eds.), Logical Foundations of Computer Science. XI, 513 pages. 2007.

Vol. 4513: M. Fischetti, D.P. Williamson (Eds.), Integer Programming and Combinatorial Optimization. IX, 500 pages. 2007.

Vol. 4511: C. Conati, K. McCoy, G. Paliouras (Eds.), User Modeling 2007. XVI, 487 pages. 2007. (Sublibrary LNAI).

Vol. 4510: P. Van Hentenryck, L. Wolsey (Eds.), Integration of AI and OR Techniques in Constraint Programming for Combinatorial Optimization Problems. X, 391 pages. 2007.

Vol. 4509: Z. Kobti, D. Wu (Eds.), Advances in Artificial Intelligence. XII, 552 pages. 2007. (Sublibrary LNAI).

Vol. 4508: M.-Y. Kao, X.-Y. Li (Eds.), Algorithmic Aspects in Information and Management. VIII, 428 pages. 2007.

Vol. 4507: F. Sandoval, A. Prieto, J. Cabestany, M. Graña (Eds.), Computational and Ambient Intelligence. XXVI, 1167 pages. 2007.

Vol. 4506: D. Zeng, I. Gotham, K. Komatsu, C. Lynch, M. Thurmond, D. Madigan, B. Lober, J. Kvach, H. Chen (Eds.), Intelligence and Security Informatics: Biosurveillance. XI, 234 pages. 2007.

Vol. 4505: G. Dong, X. Lin, W. Wang, Y. Yang, J.X. Yu (Eds.), Advances in Data and Web Management. XXII, 896 pages. 2007.

Vol. 4504: J. Huang, R. Kowalczyk, Z. Maamar, D. Martin, I. Müller, S. Stoutenburg, K.P. Sycara (Eds.), Service-Oriented Computing: Agents, Semantics, and Engineering. X, 175 pages. 2007.

Vol. 4501: J. Marques-Silva, K.A. Sakallah (Eds.), Theory and Applications of Satisfiability Testing – SAT 2007. XI, 384 pages. 2007.

Vol. 4500: N. Streitz, A. Kameas, I. Mavrommati (Eds.), The Disappearing Computer. XVIII, 304 pages. 2007.

Vol. 4499: Y.Q. Shi (Ed.), Transactions on Data Hiding and Multimedia Security II. IX, 117 pages. 2007.

Vol. 4498: N. Abdennahder, F. Kordon (Eds.), Reliable Software Technologies – Ada Europe 2007. XII, 247 pages. 2007.

Vol. 4497: S.B. Cooper, B. Löwe, A. Sorbi (Eds.), Computation and Logic in the Real World. XVIII, 826 pages. 2007.

Vol. 4496: N.T. Nguyen, A. Grzech, R.J. Howlett, L.C. Jain (Eds.), Agent and Multi-Agent Systems: Technologies and Applications. XXI, 1046 pages. 2007. (Sublibrary LNAI).

Vol. 4495: J. Krogstie, A. Opdahl, G. Sindre (Eds.), Advanced Information Systems Engineering. XVI, 606 pages. 2007.

Vol. 4494: H. Jin, O.F. Rana, Y. Pan, V.K. Prasanna (Eds.), Algorithms and Architectures for Parallel Processing. XIV, 508 pages. 2007.

Vol. 4493: D. Liu, S. Fei, Z. Hou, H. Zhang, C. Sun (Eds.), Advances in Neural Networks – ISNN 2007, Part III. XXVI, 1215 pages. 2007.

Vol. 4492: D. Liu, S. Fei, Z. Hou, H. Zhang, C. Sun (Eds.), Advances in Neural Networks – ISNN 2007, Part II. XXVII, 1321 pages. 2007.

Vol. 4491: D. Liu, S. Fei, Z.-G. Hou, H. Zhang, C. Sun (Eds.), Advances in Neural Networks – ISNN 2007, Part I. LIV, 1365 pages. 2007.

Vol. 4490: Y. Shi, G.D. van Albada, J. Dongarra, P.M.A. Sloot (Eds.), Computational Science – ICCS 2007, Part IV. XXXVII, 1211 pages. 2007.